Law, Governance and Technology Series

Volume 52

The *Law, Governance and Technology Series* is intended to attract manuscripts arising from an interdisciplinary approach in law, artificial intelligence and information technologies. The idea is to bridge the gap between research in IT law and IT-applications for lawyers developing a unifying techno-legal perspective. The series will welcome proposals that have a fairly specific focus on problems or projects that will lead to innovative research charting the course for new interdisciplinary developments in law, legal theory, and law and society research as well as in computer technologies, artificial intelligence and cognitive sciences. In broad strokes, manuscripts for this series may be mainly located in the fields of the Internet law (data protection, intellectual property, Internet rights, etc.), Computational models of the legal contents and legal reasoning, Legal Information Retrieval, Electronic Data Discovery, Collaborative Tools (e.g. Online Dispute Resolution platforms), Metadata and XML Technologies (for Semantic Web Services), Technologies in Courtrooms and Judicial Offices (E-Court), Technologies for Governments and Administrations (E-Government), Legal Multimedia, and Legal Electronic Institutions (Multi-Agent Systems and Artificial Societies).

Salvatore Sapienza

Big Data, Algorithms and Food Safety

A Legal and Ethical Approach to Data Ownership and Data Governance

Springer

Salvatore Sapienza
Bologna, Italy

ISSN 2352-1902 ISSN 2352-1910 (electronic)
Law, Governance and Technology Series
ISBN 978-3-031-09366-1 ISBN 978-3-031-09367-8 (eBook)
https://doi.org/10.1007/978-3-031-09367-8

This Springer imprint is published by the registered company Springer Nature Switzerland AG
The registered company address is: Gewerbestrasse 11, 6330 Cham, Switzerland

二千年後の君へ
To You, in 2000 Years

Acknowledgements

Humans cannot create anything out of nothingness. This book is no exception. It concludes a journey and opens a new chapter in my academic life. I could not have done it without the support of many individuals within the academic community and outside. I am greatly indebted to my scientific supervisor and mentor, Professor Monica Palmirani, for advice at the different stages of my research. She offered me her cross-disciplinary background to make me focus on the many "souls" of this book.

The Law, Science & Technology LAST-JD board professors offered me precious advice during my PhD. Among them, Professor Anton Vedder and Professor Marc Cole (also PhD supervisor) provided useful comments in the drafting of the core parts of this book. Reviewers devoted time to read this book, and for that, I am thankful to them, in particular for their insightful comments. My colleagues Giorgia Bincoletto, Chantal Bomprezzi, Federico Galli and Valentina Leone supported me with frequent exchange of opinions during our doctorate and sincere friendship also beyond what was required of them as co-workers (especially abroad). Noemi Conditi, Rachele Carli, Pietro Dunn and Paola Salomone, the newcomers in my PhD programme, made my doc-to-post-doc transition smooth and lovely. The last round of PhD students, Marco Billi, Gabriele Suffia, Malwina Wojcik and Yasaman Yousefi, are currently contributing to enhance my views with their new and fresh perspectives. Finally, not strictly related to the LAST-JD, yet linked to the CIRSFID—ALMA-AI research centre at the University of Bologna, I am thankful to Alessandra Foschi, Dino Girardi Jacopo Patrignagni, Vittoria Pistone and Michela Rossi for having provided the perfect working environment. My informatics colleagues—Leticia Decker de Sousa, Davide Liga and Francesco Sovrano—contributed to improve my knowledge on technical topics. Outside of the LAST-JD circle, other people contributed to shape my research personality with constructive suggestions. The group made by Stefano Quintarelli, Francesco Corea, Fabio Fossa and Andrea Loreggia challenged my most monolithic views by means of their heterogeneous expertise.

Finally, family and friends. Even though they can read English, my guess is that they would like to read some thoughts in our common language. Ringrazio moltissimo i miei genitori e il resto della mia famiglia per il supporto ricevuto in questi anni. Se sono arrivato a questo mio traguardo è anche grazie alla serenità che hanno saputo garantire. Tra i miei preziosi amici ringrazio 353431541224344543 volte Andrea (X2), Sergio, Angelo, Gabriele, Giovanni, Giorgio (X2), Carlo, Ferdi, Agnese e Chiara. In ordine sparso ringrazio anche Giulia, Edoardo, Mattia, Alberto, Gianluca, Maddalena, Carlo, Vittorio, e tutte le altre persone che ho dimenticato ma che meriterebbero a pieno titolo di essere qui.

Contents

Acronyms

ADM	Automated Decision-Making
AI	Artificial Intelligence
ALTAI	The Assessment List on Trustworthy Artificial Intelligence
ANN	Artificial Neural Network
API	Application Programming Interface
BfR	Bundesinstitut für Risikobewertung
BSE	Bovine Spongiform Encephalopathy
CAPI	Computer-Assisted Personal Interview
CATI	Computer-Assisted Telephone Interview
CJEU	Court of Justice of the European Union
CPR	Common-Pool Resource
CSV	Comma-Separated Value
DCF	Data Collection Framework
DRM	Digital Rights Management
DTM	Document-Term Matrix
EC	European Community
ECJ	European Court of Justice
EFSA	European Food Safety Authority
ERI	Emerging Risk Identification
ERIS	Emerging Risk Identification System
EU	European Union
EUCFR	European Charter of Fundamental Rights
FAO	Food and Agriculture Organisation of the United Nations
FDA	Food and Drugs Administration
GDP	Gross Domestic Product
GDPR	General Data Protection Regulation
GFLR	General Food Law Regulation
GLP	Good Laboratory Practice
GMO	Genetically Modified Organism
HACCP	Hazard Analysis and Critical Control Points
HIC	Human In-Command

HITL	Human In-The-Loop
HLEG	European Commission High-Level Expert Group
HOTL	Human On-The-Loop
IARC	International Agency for Research on Cancer
ICESCR	International Covenant on Economic, Social and Cultural Rights
ICO	Information Commissioner's Office
ICT	Information & Communication Technology
IEC	International Electrotechnical Commission
IEEE	Institute of Electrical and Electronics Engineers
IMSOC	Information Processing System for Official Controls
IOT	Internet of Things
IPR	Intellectual Property Right
ISO	International Organization for Standardization
LoA	Level of Abstraction
MIKE	Method for an Integrated Knowledge Environment
NGO	Non-Governmental Organisation
NLP	Natural Language Processing
OCA	Principal Component Analysis
PAN	Pesticide Action Network
PPP	Plant Protection Products
QPS	Qualified Presumption of Safety
R&D	Research and Development
RALC	Restricted Access/Limited Control
RAR	Renewal Assessment Report
RASFF	Rapid Alert System for Food and Feed
RFID	Radio-Frequency Identifications
RPC	Raw-Primary Commodity
RRR	Reduced Rank Regression
SME	Small and Medium-sized Enterprise
SQL	Structured Query Language
SVM	Support Vector Machine
TDM	Term-Document Matrix
TFEU	Treaty of Functioning of the European Union
TF-IDF	Term-Frequency /Inverse Document Frequency
TRIPS	Trade Related Aspects of Intellectual Property Rights
UK	United Kingdom
WHO	World Health Organization
WSN	Wireless Sensor Networks
XML	eXtensible Markup Language

Chapter 1
Food, Big Data, Artificial Intelligence

1.1 Emerging Trends in Food Technology

On 13 October AD 54, Emperor Claudius suddenly died. Historians such as Pliny the Elder, Suetonius, and Tacitus described his death as mysterious, yet certainly linked to the ingestion of some foods, most likely mushrooms. One of their reconstructions focuses on the role played by Halotus, Claudius' personal food taster. It is likely that Halotus was involved in a conspiracy lead by Agrippina to bring Nero to the throne, but there was not enough evidence to convict him.[1] More recently, Adolf Hitler, Vladimir Putin, US Presidents including Barack Obama have reportedly hired food tasters to prevent food poisoning.[2]

Common people do not rely on food tasters to check the quality of their food. We make some intuitive controls right before consumption by our senses (e.g., sight, smell) and rely on our past experience to understand whether or not our food is safe to be eaten. However, with the recent developments in biochemistry, the globalisation of food market, and the elongation of the food chain, we feel powerless when wondering whether certain products are safe for our health, in particular as regards long-term effects.

[1] Grimm-Samuel (1991)

[2] See, for instance, in the press "One Bite Away from Death", available at https://www.spiegel.de/international/germany/hitler-food-taster-margot-woelk-speaks-about-her-memories-a-892097.html, accessed on 27/04/2022; "Vladimir Putin employs a full-time food taster to ensure his meals aren't poisoned" https://www.independent.co.uk/news/world/europe/vladimir-putin-employs-a-fulltime-food-taster-to-ensure-his-meals-aren-t-poisoned-9624380.html, accessed on 27/04/2022; "Yes, Obama Probably Has A Food Taster—But So Did Ronald Reagan" https://www.businessinsider.com/plenty-of-evidence-president-food-taster-2013-3?r=US&IR=T, accessed on 27/04/2022

S. Sapienza, *Big Data, Algorithms and Food Safety*, Law, Governance and Technology Series 52, https://doi.org/10.1007/978-3-031-09367-8_1

When these trends became popular in the 1990s, a parallel development of food safety agencies occurred in the European Community. Thus, we started to delegate all the 'hard' risk assessment activities to the scientists working for these authorities. Their output consisted of scientific opinions and administrative decisions (e.g., authorisation, banning, recalls, etc.) were taken on their grounds. Therefore, even though we cannot always be sure that our food is safe, we trust that the food safety system has worked correctly by preventing negative effects on our health. Misplaced trust, as with the case of Claudius and Halotus, is extremely negative also today: alongside fatalities or injuries, the consequences of 'food scandals', which will be presented, are negative for the whole system, including the food industry.

Today, recent technological trends are progressively reshaping the way in which food safety risk assessment is carried out and decisions are taken. Data availability and computational power make possible the analysis of large, heterogeneous, up-to-date information, also by automated means. This fascinating evolution entails risks and opportunities that are necessary to be evaluated. Consistently with other domains affected by similar traits, it is necessary to steer the adoption of such novels forms of analysis in a way that maximises the opportunities while minimising the risks. Our agencies can do a better job in evaluating food hazards. The question remains how.

However, differently from other research domains in data governance, data protection law and ethics, little or no attention has been devoted to food safety. The reasons might be multiple. It might be because such transition has still to be completed, or because the technical and legal frameworks—as they will be shown—are rather complex, or because COVID-19 pandemics have steered the attention of researchers towards more actual topics in the area of public health.

This **Introduction** explains why food safety and its emerging trends deserve attention. Novel technical trends raise serious questions regarding the governance of new technologies in the food safety domain. This is possibly due to the presence of heterogeneous stakeholders whose interests are not always compatible. Such contrasts hinder shared solutions on the direction to take.

Chapter 2 discusses the state of the current debate in data ownership, which represents a crucial issue in this research domain. Data are valuable assets for the industry, but they are also our gateway to see if our food safety authorities are taking the right decisions. The discussion is carried out also in the light of recent legislative initiatives and cases that have brought new light to this topic.

Chapter 3 is dedicated to a particular information-type that is peculiar in food safety risk assessment, i.e., food consumption data. This information is necessary to evaluate the existence of possible hazards to human health. However, its legal status is contested and governance solutions are needed to minimise the risks to fundamental rights and freedoms associated with its processing. The relevance of these data in data protection law, their proximity to personal identity, and their relevance in a group-oriented discourse are adopted as a key approach.

Chapter 4 explores open questions in food safety data governance. This wide range of topics covers *inter alia* well-known topics in "Artificial Intelligence governance" literature, such as algorithmic transparency, fairness in risk assessment, and explainability. Moreover, it explores a piece of legislation currently under discussion, i.e., the EU Data Governance Act.

Chapter 5 proposes a unified ethical approach, the P-SAFETY model, which constitutes the most original contribution of this book to the current debate. A list of principles, identified and discussed within the current debate, is proposed to tackle old and emerging issues in this research domain. Further, its rationale and some possible implementations are also discussed.

Finally, **Chap. 6** summarises the main contents of this books and sets directions for future works.

1.1.1 Big Data, the Food Industry and the UN Sustainable Development Goals

The 2010s have seen the consolidation of "Big Data" trends in several sectors, including healthcare, banking and financial sectors, automotive,[3] and a significant growth of related research fields.[4] The seminal work by Mayer-Schönberger and Cukier made clear how the impact of the rise of a data-driven economy is largely based on the availability of digital information.[5] Big Data remains an ambiguous term. Mittlestadt and Floridi affirm that there is no unique understanding of the concept of Big Data in philosophical terms,[6] nor it has been found by extensive reviews of the technical literature on the topic.[7]

Back in 2001, Douglas Laney proposed a tripartition of Big Data into *Volume*, *Velocity* and *Variety*:[8] Volume identifies the dimensional aspect of Big Data; Velocity is related to the update frequency of this information; Variety describes the presence of heterogeneous formats (e.g., photos, web pages, personal information, audiovisual) that can combined in a single operative framework and then linked by specific formats, such as eXtensible Mark-up Language (XML). Taking inspiration from Laney's work, others expanded his definition by adding further "V"s. IBM first included "*Veracity*", i.e., the necessity of having high quality data to prevent false results. Then, the American company added a further "V" by identifying "*Value*" as the economical component of Big Data.[9] A neighbour definition is also part of the ISO/IEC 20546 on the concept of Big Data (para. 3.1.2), which also takes into account the technology used to manage extensive databases. The *V*-based conceptualisation has become paradigmatic and widely accepted.

[3] See, respectively, Luckow et al. (2015); McAfee et al. (2012); Morabito (2015)

[4] Chen et al. (2014)

[5] See Mayer-Schönberger and Cukier (2013), in particular Chapter 6 and Chapter 7

[6] Mittelstadt and Floridi (2016)

[7] *Inter alia*, De Mauro et al. (2015) and Curry (2016)

[8] Laney (2001)

[9] IBM (2014, 2016)

Several correlated factors have led to the establishment of a data-driven economy: the widespread adoption of digital devices that serve as data sources, lower costs in data storage, and sufficient computational power to "make sense" of this information jointly contribute to the rise of 'Big Data' applications.[10] Despite the data-centric nature of their definition, 'Big Data' as a consolidated trend does not only correlate to bulk information collection, but it is also linked to a more dynamic perspective of potential exploitation through data analysis or 'mining'.[11] This wording mirrors the transformative uses of data (from raw materials to information) discussed in economic theory,[12] i.e., according to categories strictly related to the economic value—rather than the epistemic significance—gained from information gathering and processing.

Machine Learning is currently one of the most popular approaches to data analysis.[13] Placed within the broader category of Artificial Intelligence (AI),[14] Machine Learning has developed under two approaches, i.e. the Symbolist and the Connectionist,[15] which have historically driven the research on AI towards symbolic and subsymbolic paradigms. Symbolic approaches to AI attempt to process input symbols according to pre-defined explicit rules which mimic human's mind activity in order to produce an output. They consist of logic-based (i.e., based on logically-constructed rules) or knowledge-based (i.e., based on ontologies) *a-priori* rules. Instead, subsymbolic approaches, do not provide any *a priori* knowledge to follow for computer reasoning. It will further discussed that this comes at the cost of explaining the reasoning of the algorithm and its results.[16]

Machine Learning is the subset of AI consisting of computational methods in which 'algorithms are trained to infer certain patterns based on a set of data in order to determine the actions needed to achieve a given goal'.[17] Machine learning algorithms can be broadly classified according to the following categories:

- **Supervised learning** aims to find patterns between labelled variables/predictors to infer the value of a target variable. Hence, the generation of a statistical

---

[10] Manyika et al. (2011)

[11] Corea (2016)

[12] Moody and Walsh (1999)

[13] Müller and Guido (2016)

[14] The definition of Artificial Intelligence is still under discussion. For an overview of the topic, see the seminal work by Russell and Norvig (2010), and the discussion in HLEG (2019). A legal definition of AI systems—further discussed in Chap. 5—is the one proposed by the draft EU Artificial Intelligence Act: "'artificial intelligence system' (AI system) means software that is developed with one or more of the techniques and approaches listed in Annex I and can, for a given set of human-defined objectives, generate outputs such as content, predictions, recommendations, or decisions influencing the environments they interact with"

[15] Russell and Norvig (2010), p. 24

[16] Pasquale (2015)

[17] European Commission (2020), p.16. The paternity of the expression 'machine learning' is attributed to Samuel (1967). For an overview on the definitions of ML see Corea (2019), HLEG (2019), Mitchell et al. (1997), and Jordan and Mitchell (2015)

supervised model requires a manually annotated or labelled training dataset that is used to train an algorithm to find patterns (statistical regularities) in data.

- **Unsupervised learning** aims to find patterns between data points to organise and represent them without the need of labelled variables;
- **Semi-supervised learning** consists of a hybrid method needed when a limited amount of labelled data is available;
- **Reinforcement learning** makes algorithms choose an action based on the environment in which they are deployed. Eventually, they get penalised or rewarded on the basis of their actions;
- **Ensemble methods** consist of the combination of two or more approaches.

The food chain is no exception to 'Big Data' and 'data analytics' trends. The availability of large quantities of real-time data and the possibility to extract valuable information from them has led to a substantial growth in the investments made to transform the food industry into a data-driven economy. Consultancy firm Accenture reports that '[a]dvances in technology and open access to capital will fuel an explosion of business model innovation'.[18] Agriculture is undergoing a digital revolution, while supply chain management is empowered by data-driven tracking strategies.[19] Consumers become more aware of nutritional facts and increasingly track their food behaviour via mobile 'fitness' applications which might pave the way to nutritional research.[20] From a broader perspective, COVID-19 pandemic and the 'stay at home' preventive response has increased the market shares of food delivery services, as case studies in the United States, Taiwan, Germany, Spain, the Netherlands and Italy[21] suggest. Since delivery platforms usually operate via websites or mobile applications, data about food preferences are generated and made available for analysis in larger quantities than pre-COVID era.

Beyond their obvious link to economics, technical advancements across the food chain have relevant social implications. Of all the human activities, eating is one of the most necessary for our survival and it is linked to our health. Therefore, foodstuffs are subject to a close scrutiny by national, regional and supranational authorities, which often regulate their production, distribution, labelling, packaging, and marketing.

In 1974, the Food and Agriculture Organisation of the United Nations (FAO) World Food Summit described food security as the commitment to ensure "availability at all times of adequate world food supplies of basic foodstuffs to sustain a steady expansion of food consumption and to offset fluctuations in production and prices". While this first definition was mainly concerned with availability, and the stability of times and prices of food supplies, a subsequent amendment included the concept of "access to food" in the definition of food security. Eventually, the mid-80s definition

[18] See the report by Accenture (2017)

[19] Bronson and Knezevic (2016) and Govindan et al. (2018)

[20] Wolfert et al. (2017), Murnane et al. (2015), and Maringer et al. (2018)

[21] See, respectively for each Country, Gavilan et al. (2021), Chang and Meyerhoefer (2021), Dannenberg et al. (2020), Laguna et al. (2020), Poelman et al. (2021), and Alaimo et al. (2020)

provided by the Food and Agriculture Organisation (FAO) conceptualised food security as the goal to ensure "that all people at all times have both physical and economic access to the basic food that they need".[22]

It may be argued that the stable availability and the access to sufficient food are the only necessary conditions to achieve food security, but this is not the case in reality. Accessibility to food alone might be insufficient to guaranteeing an healthy life. Following these considerations, in 1986 the World Bank defined food security as the "access of all people at all times to enough food for an active, healthy life". Such teleological clarification was needed to broaden the scope of food security in order to include the reason *why* States should be committed to ensure food security, i.e., protecting citizens' health. Within European Union (EU) Member States, where most people do not experience undernourishment[23] these qualitative approaches to food security mainly aimed at promoting policies that guaranteed safe food, only after the occurrence of "food crises" which triggered the attention of lawmakers and paved the way for 'heavy hand' forms of regulation[24] that will be discussed throughout this book (in particular, in Chap. 2).

A notable example of foodborne outbreak is the Bovine Spongiform Encephalopathy (BSE)—commonly referred to as the Mad Cow disease—first reported in 1986.[25] The outbreak is considered one of the most significant and harmful food-borne epidemics of recent history in Europe. Once eradicated, the numbers describing this crisis were unforgiving: in the United Kingdom only, 180,000 cattle were infected and more than four million cattle were slaughtered. It is reported that 178 people died from mad cow-related diseases.[26] Similarly, the 2011 Escherichia Coli O104:H4 bacteria outbreak—as reported by the European Food Safety Authority (EFSA)—severely hit Germany and neighbouring countries, resulting in 4026 cases, of which 51 were fatal.[27] Another more recent incident, the Salmonella outbreaks of 2012, is deemed to be responsible for 65,317 cases and 61 deaths in the EU.[28]

Taken together, the starvation that is severely affecting several regions worldwide and the concerns for food safety call for an international commitment in the UN Sustainable Development Goals (SDGs). Goal 2: "Zero Hunger" and Goal 3: "Good Health and Well-being" are two key areas in which advancements in food technologies, including Big Data and advanced data analytics techniques, can have a significant impact. This is consistent with the claims raised by scholars who argued about socially beneficial outcomes made possible by AI applications deployment in the context of UN SDGs.[29] Therefore, legitimate and significant expectations

[22] FAO (1983)

[23] FAO (2018)

[24] Alemanno and Gabbi (2016) and Knowles et al. (2007)

[25] Brown et al. (2001)

[26] In the news, BBC (2018)

[27] EFSA (2011)

[28] EFSA (2014)

are placed on food-related technological advancements that rely on Big Data, inferences, and predictions. International actors, including the World Economic Forum, claim that with the growth of global population predicted for the upcoming decades, AI will contribute to feed the hungry while preserving biodiversity and sustainability.[30]

Despite assumptions and claims about the potential of AI technologies, several factors contribute to adjust the expectations towards machine learning. *Inter alia*, epistemic and normative concerns have been raised by scholars investigating the impact of Big Data and algorithms from an ethical perspective.[31] Within the epistemic set of moral issues, these scholars have argued that inconclusive, inscrutable and misguided evidence gathered from the use of machine learning systems can lead to unjustified, opaque and biased actions. In turn, they can result in unfair and discriminatory outcomes or morally-questionable reontologisations of the reality which threaten informational privacy (e.g., profiling), that are included amongst normative concerns. Finally, traceability of algorithms consists of a unique set of problems linked to the intrinsic opaqueness of certain machine learning computation that result in accountability gaps. Despite the significant increase in research towards such ethical concerns,[32] also in 'neighbour' fields (e.g., healthcare[33]), no study has discussed their implications and significance food-related matters.

1.1.2 Sensors and Traceability in Agriculture 4.0

As noted in the previous section, the impact of AI systems in industry is leading to a significant increase in investments towards the digitalisation of industrial sectors.

One of the key areas of technical advancements is agriculture. Being one of the most relevant subsets of the primary sector, farming is highly influenced by ICT-driven transformations. Reviewers[34] have identified several ongoing trends in agriculture 4.0, also known as precision agriculture (PA) or smart farming. These trends are not characterised by the use of a single technology, but they are made possible by the combination of multiple ICTs aimed at improving the profitability and sustainability of farming, e.g., by increasing the degree of automation in certain tasks or by improving decision systems.[35] Despite being dated back to the 1980s,[36]

---

[29] Cowls et al. (2021)

[30] World Economic Forum (2020)

[31] Mittelstadt et al. (2016)

[32] Morley et al. (2019)

[33] Morley et al. (2020)

[34] Wolfert et al. (2017) and Miles (2019)

[35] Es et al. (2016) and Sonka (2014)

[36] Mulla (2013)

PA is now scalable due to lower costs in microprocessors and the implication of new technologies such as cloud computing or mobile applications.

Geographic Information Systems (GISs) combine spatial data with soil information.[37] This technology is not entirely new since it was originally proposed by FAO in the 1990s, but its deployment has been recently proved to be successful in several case studies discussed by the literature in the fields of urban extension, deforestation, and climate change.[38] Soil sensors aggregate these data with satellite images.[39] Recent data acquisition trends also relate to the use of Unmanned Arial and Ground Vehicles (UAVs and UGVs respectively).[40]

The deployment of Internet of Things (IoT) technologies—smart objects connected via each other through a wireless network infrastructure[41]—has been fruitful in crop and resource management and monitoring, with increases in quality and quantity of the crop yield.[42] Data acquisition is also related to environmental information, such as humidity, temperature, and light.[43]

The proliferation of mobile applications for smart farming has also been observed by reviewers,[44] also in developing countries.[45] Mobile applications allow data access and visualisation, thus contributing to the decision-making processes regarding crop nutrition, fertilisation and irrigation. Smartphone cameras are used for crop protection and diagnosis purposes.

Blockchain technologies are deemed to be game-changing for traceability across the whole food chain.[46] Use cases conducted by IBM and Walmart and Carrefour show the potential of this technology in food logistics.[47] In smart farming, blockchain can be used in the early stages of the food chain by the seed provider the producer.[48] Given the presence of several actors involved in the agricultural supply chain and the international nature of the trades, blockchain can reduce the time required for manual checks and document processing.[49]

The adoption of smart farming approaches depends on different factors such as the size of the farm, the location, the soil quality, socio-demographic factors (including digital literacy) and on financial capacity.[50] Key drivers have attracted

---

[37] Lytos et al. (2020)
[38] Rossiter (1990) and Bielecka (2020)
[39] Balafoutis et al. (2017)
[40] Vu et al. (2018)
[41] Porter and Heppelmann (2014)
[42] Dagar et al. (2018)
[43] Balafoutis et al. (2017)
[44] *Ex multis*, Mendes et al. (2020)
[45] See Baumüller (2017)
[46] Tian (2016)
[47] See Kamath (2018) and Carrefour (2018)
[48] Kamilaris et al. (2019)
[49] Hoffman and Munsterman (2018)
[50] Pierpaoli et al. (2013)

the attention of the EU Parliament.[51] Crucially, the deployment of "smart farming" application is taking place in Europe and North America, whereas China is leading the research on this topic.[52] It is reported that, depending on the crop, the adoption of smart farming solutions can double profitability per acre.[53]

Taken together, the emergence of smart farming technologies is not exempt from significant issues, ranging from the competitive power gained by the deployment of ICTs, to the ownership of the data collected and processed, to the privacy of farmers and other people involved in related activities, and to the epistemic aspects of the decisions taken on the basis of suggestions of systems such as mobile applications. For instance, the question of data ownership was briefly introduced by a study conducted by the EU Parliament,[54] whereas other aspects have been discussed by the academic literature,[55] however, in atomistic fashions that do not approach the subject matter holistically. Integrating this knowledge will be one of the goals in the next chapters.

1.1.3 The in silico Age of Food Safety Risk Assessment: From Food Tasters to Machine Learning

Europe does not experience hunger. However, it is highly concerned by food *safety* (rather than *security*). Promising advances in agriculture and in the food chain technology might suggest that the evaluations of risks connected to foods can be fostered by the significant amount of data generated and processed across the chain, including those related to consumer preferences (e.g., mobile applications).

Originally, tasting *all* the foods right before consumption was the only form of risk assessment in Ancient Rome. "Food tasters" were often appointed by rule-makers—nobles, kings and emperors—to prevent food poisoning, an easy way widely used to kill opponent. However, as explained above, the employment of food tasters did not prevent occasional murders, as the example of Emperor Claudius killed by his own food taster taught us.

In the future, the increasing use of data to classify risks and generate predictions will de-materialise food safety risk assessment. Throughout this book, such trend will be referred to as an ongoing "datafication"[56] process, whose peculiarities and implications will be progressively identified. It can be preliminarily observed that this trend has two main facets: *ontologically* speaking, the concept of food and associated entities (consumption, genetic modifications, placement on the market,

[51] See the report by the European Parliament Research Service, EPRS (2016)

[52] Faulkner et al. (2014) and Wolfert et al. (2017)

[53] Accenture (2020)

[54] See the report by the European Parliament Research Service, EPRS (2016), p. 9

[55] Leone (2017)

[56] See Mayer-Schönberger and Cukier (2013)

and so on) is translated into computable objects; in the *epistemic* perspective, the computability of food and the same components entails that ICTs can be used to understand the implications of food-related behaviours and draw machine-supported or machine-generated inferences. A joint reading of this twofold perspective might suggest that, from a purely technical point of view, the ongoing datafication process consists of an abstraction of foods and their components into machine-readable information on which humans test and validate scientific hypotheses.

However, this abstraction does not imply that a completely "synthetic" form of risk assessment that is detached from tangible food and samples would be possible. Even in the most futuristic and "datafied" scenario, foodstuffs will likely remain necessary components of safety risk assessments since they are the ontological and epistemic centre of the analysis. Risk assessment shall also be conceived as an epistemic activity and, as such, it implies some degree of *understanding* in the sense given by Durante.[57]

Nonetheless, the presence of artificial agents that cooperate with scientists to predict trends in outbreaks or to identify contingent risks is likely to become reality, seemingly to what happens in the biomedical context.[58] The great promise of this transition, aligned with UN SDGs and remote risk assessment, is that everyone on Earth will benefit from such advances due to the better understanding of the peculiarities of each risk by targeted analysis and integration with raw data from individuals, fields, industrial actors.[59]

Today, even though the datafication of risk assessment is tangible and algorithmic evaluation methods are rapidly developing, microbiology and traditional methodologies are still the dominant research paradigm (Fig. 1.1). As we live in the "mangrove society",[60] technological advances are shaping a domain in which the relevance of data is becoming equal to—if not higher than—the one of the observation of real and tangible food. In the course of its de-materialisation, datafied and data-centric food safety generates opportunities and risks.

---

[57] See Durante (2019), p.192. As he argues, understanding is an exclusive capacity of humans even though we are progressively adapting the environment to make it suitable for artificial agents. See also Floridi (2014), ch.2 and Durante (2019), p.236 to which we delegate some epistemic-related tasks. However, the activity of *understanding* cannot be delegated to non-human agents

[58] Morley et al. (2020)

[59] The deployment of modern computational techniques based on large quantities of up-to-date data, AI and ML is only of the factors that might contribute to make a timely and accurate food safety risk assessment available to everyone. Other factors include the role of international cooperation, the industry, education, public policies, and so forth. This does not necessarily imply that any discussion towards Big Data, AI and ML is nonsense. Critical scenarios might emerge from their deployment, e.g., the increase of the gap between developed and developing countries in terms of food safety, or environmental issues linked to data-driven practices related to regulated products, including animal life. However, this book will dedicate most of its attention to issues that specifically pertain to data, algorithms, and practices related to food safety and, in particular, risk assessment

[60] Floridi (2018)

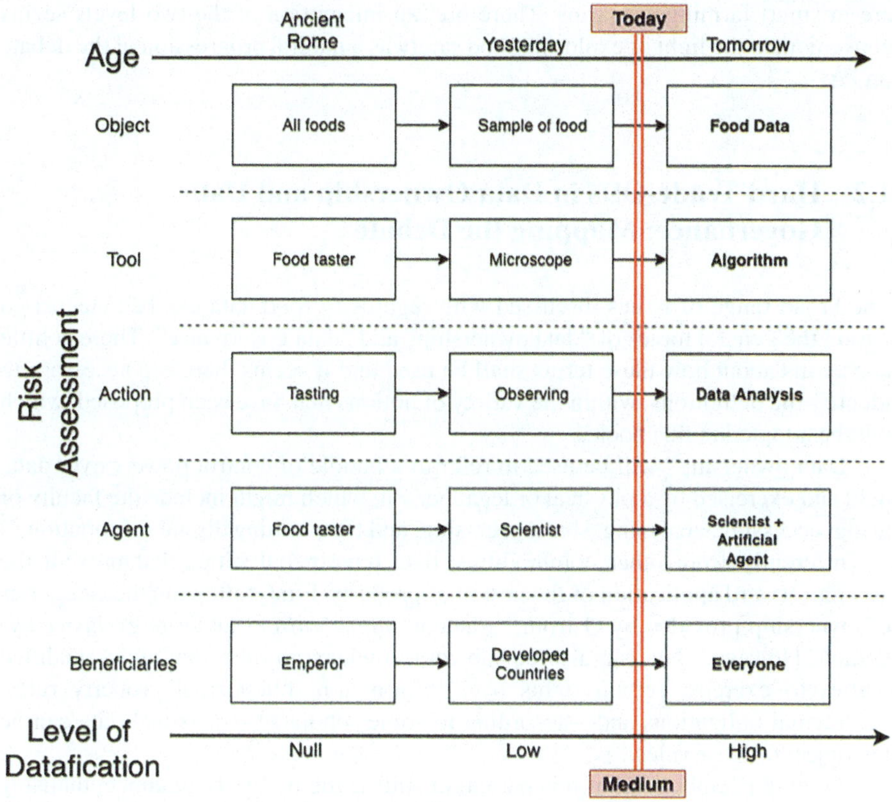

Fig. 1.1 Past, present and future of food safety risk assessment

Several reasons can be attributed to the rise of computational approaches to risk assessment. Similarly to other domains, they might include the availability of larger quantities of structured and updated information—usually referred to as "Big Data"—and sufficient computational power to analyse them in novel forms, including machine learning.[61] This trend is not specific to food safety, but this domain is undoubtedly one of the least discussed. While commentators have already noted that digital technologies have the potential to enhance the interoperability of data, hyper-linking knowledge, performing aggregate analysis and visualisation also in food safety,[62] a detailed study of these phenomena and their social and legal implications is still missing.

Moreover, issues pertaining to competition, ownership of data, privacy and algorithmic decision-making are still relevant in the food safety domain as they

---

[61] Floridi (2014), Ch.3

[62] Alemanno (2014), p.213

are in smart farming scenarios. Therefore, an integration of the two levels seems convenient in the light of exploring food safety as a natural progression of the debate on PA.

1.2 Hard Trade-Offs in Data Ownership and Data Governance: Mapping the Debate

The broad range of issues discussed with regards to food data can be categorised within the general topics of "data ownership" and "data governance". There is little consensus about how these terms shall be used and it seems therefore necessary to identify the definitions, within the variety of notions that have been proposed, which will be adopted in this book.

"Data ownership" will be used to refer to a bundle of control powers over data, held and exercised by a physical or legal person, which might include the faculty of using, accessing, analysing, sharing, erasing, and transmitting digital information.[63]

Differently from other commodities, data have peculiarities that prevent the straightforward application of the notion of property. First, following the categories of ownership provided by Ostrom,[64] data are *non-rivalrous* and *non-exclusive* by-default. However, data can also be subject to ownership rules that allow qualified parties to exercise certain rights over information. Intellectual property rules, contractual obligations and—according to some scholars[65]—personal data can be the object of these rules.

The major implication of this paradigm shift is the necessity of conceptualising data ownership as a scalar measure. It does not consist of "all or nothing" prerogatives like the traditional and monolithic *ius utendi, fruendi et abutendi* concept of property derived from Roman Law,[66] but it can be conceptualised as a bundle of rights (e.g., access, distribution, etc.) related to informational assets that show some degree of control over the information. This also implies that such control tends not to be exclusively granted to solely one physical or legal person, as multiple ownership models can co-exist on the same informational assets. Let us consider the following scenario: a copy of the Divine Comedy is freely accessible online on an open folder stored in a cloud service. While that file remains public (a), someone could download it and protect her or his copy with a password (b),

[63] An extensive discussion on this definition will be provided in Sect. 2.1.1

[64] Ostrom et al. (1990)

[65] See, *inter alia*, the points raised in Malgieri (2016), or the ownership-based rationales for personal data protection in Tamò-Larrieux (2018)

[66] In 1569, French humanist François Hotman described property (*Dominium*) as the "Ius ac potestas re quapiam tum utendi, tum abutendi, quatenus iure civili permittitur", i.e. the right to use and consume anything, as permetted by the civil law. The definition broadens the one previously given by Bartolo and became influential for the drafting of the Code Napoleon. See Moréteau et al. (2019)

whereas a teacher could download it and upload it on a school repository accessible only to her or his students (c) or in a private server which can be accessed only by a limited number of individuals at the same time (d). The same exact digital copy falls within all the ownership models—public good (a), private good (b), club good (c), common-pool resource(d)—at the same time.

Therefore, 'ownership' will be used in this book to identify certain faculties which include the power of excluding others to enjoy a given piece of information or, conversely, to grant third parties the possibility to access, share and transmit data. Our definition is agnostic with respect to the other definitions (including legal ones, e.g., 'data controller', 'data subject', 'right-holder') that can apply contextually to 'data owner'.

Two major uses of 'data governance' can be found in the literature and in common language. One is mostly associated with the idea of data "as-an-asset" held by private entities and it is heavily connected with business-oriented data analysis, storage, and use. Many different expressions describe the specific emphasis that data owners should put on data-related issues. "Data governance" is the branch of Enterprise Information Management (EIM) aiming at maximising data value by improving data quality.[67] Similarly, Data governance has been defined as "the formulation of policy to optimize, secure, and leverage information as an enterprise asset by aligning the objectives of multiple functions".[68] Synonyms also include "data quality management".[69] Data governance "models" have been drafted starting from the level of maturity of a private entity towards data and organised in steps.[70] IBM, Oracle, and Gartner are popular examples of maturity-based data governance models.[71]

The second definition of 'data governance' is less focused on the operational use of data by companies and keener on discussing "all processing of governing".[72] "Governance" derives from the Greek χυβερνάω (to steer) and this second meaning reflects the one originally conferred by Plato in his *Republic*.[73] Thus, it transcends the traditional meaning of State, law, government and can be identified as the set of decision-making processes within a community. Such processes encompass governments, markets and networks, thus being a distinctive feature of multi-stakeholder environments. 'Hard' law is not the only governance tool: hybrid practices often combine administrative systems with market strategies and no-profit arrangements.[74] Mixed forms of governance can also apply to ICT-related contexts.[75]

---

[67] Ladley (2012), p.7, p. 101

[68] See the introductory work by Soares (2015)

[69] For instance, in the meaning given by Wende (2007)

[70] The Data Warehousing Institute (2013)

[71] See, respectively, Firican (2018), Oracle (2013), and Gartner (2008)

[72] Bevir (2012)

[73] Grindle (2007)

[74] Bevir (2012)

In this second meaning, "data governance" has already been conveniently defined by several entities. A similar expression, "information governance" has been defined as "the specification of decision rights and an accountability framework to encourage desirable behaviour in the valuation, creation, storage, use, archival and deletion of information. It includes the processes, roles, standards and metrics that ensure the effective and efficient use of information in enabling an organization to achieve its goals.".[76] Then, the Data Governance Institute has affirmed that "Data Governance is a system of decision rights and accountabilities for information-related processes, executed according to agreed-upon models which describe who can take what actions with what information, and when, under what circumstances, using what methods.". [77] Moreover, data governance has also been defined as "the formal execution and enforcement of authority over the management of data and data-related assets", "formalizing behaviour around the definition, production, and usage of data to manage risk and improve quality and usability of selected data" and "formalising and guiding behaviour over the definition, production, and use of information and information-related assets"[78]

Accordingly, the UK British Academy and the Royal Society have defined "data governance" in more holistic terms by including "legal, ethical, professional and behavioural norms of conduct, conventions and practices that, taken together, govern the collection, storage, use and transfer of data".[79] This definition looks appropriate for the context of this book. Its major advantages consist of reconciling ethical norms and rules, defining certain types of data processing, and including practices among the component of data governance. For these reasons, we will adhere to this definition throughout this book.

As regards the relationship between ownership and governance, it might be useful to rely on the jurisprudential distinction between legal positions and ethical and legal values.[80] Data ownership can be expressed in terms of legal positions in Hohfeldian concepts.[81] These are legally-qualified statuses—right, priviledge, power, immunity, no-right, duty, disability, and liability—attributed to persons, to which other legal positions, attributed to other persons, correspond in correlation or in opposition.[82] Instead, data governance will refer to the way in which such data-

[75] See Pagallo et al. (2019). Chapter 5 will discuss in detail the proposed 'middle-out' approach

[76] Logan (2010)

[77] The Data Governance Institute (2015)

[78] Seiner (2014), p.2–3

[79] British Academy and Royal Society (2017)

[80] The author thanks the reviewer for the thoughtful comment and suggestion

[81] See the foundational works by Hohfeld (1917a,b) and a recent commentary by Hage and Pfordten (2009)

[82] See the example by Singer (1982): "If I "own" property, it means that I have various rights with respect to the thing constituting my property–the "bundle" of sticks or rights. I probably have the right to exclude and everyone else in the world has a correlative duty not to use my property. Some people may have a privilege, however, as to fly over it. I also have power with respect to my property because I can create rights in others, as by transfering some or all of the property to them,

related legal positions are constituted (e.g., by law or codes of conducts), preserved (e.g. through sanctionatory mechanisms), and legitimised (e.g., by relying on ethical considerations).

Once clarified the meanings of data ownership and data governance for the purposes of this book, it is possible to map the key areas of debate surrounding the relationship between food and Big Data. Section 1.2.1 below identifies the key arguments in favour of a strict or wide availability of company and environmental data, whereas Sect. 1.2.2 tackles the same issue from the perspective of personal information. Section 1.2.3 briefly introduces the interplay between Big Data and competition in the domain of food-related information.

1.2.1 Property Rights, Confidentiality and Transparency

Company data are an asset. They can offer significant competitive advantages for entities capable of collecting, organising and analysing them. Therefore, data acquisition is always paired with the necessity of protecting the investments made to gather the information and keep the inferences made away from public scrutiny. This is done by a plethora of instruments, including intellectual property. In the EU framework, The Database Directive[83] grants a *sui generis* protection for the maker of the database to prevent extraction and/or re-utilisation of data if a quantitatively/qualitatively relevant investment has been made in obtaining, verifying or presenting the contents of the database (Art. 7).

Food-related data do not constitute an exception. Companies active in this market sector are willing and entitled to protect their data. However, significant differences can be observed among player markets, depending on the specific sub-sector to which they belong. For instance, scholars have discussed agricultural data ownership under the perspective of finding the entity entitled to exploit the economic value of information[84] or in terms of the relation and distribution of power between tech companies providing digital infrastructures and farmers.[85] In both perspectives, re-thinking data ownership has been considered necessary, also based ethical on considerations.[86]

An additional layer of complexity is given by the protection of commercially sensible information related to certain regulated products, including those used in

as by creating an easement, which gives the grantee certain rights vis-a-vis others and certain rights and privileges vis-a-vis me.". See also the explanations given in the field of 'artificial intelligence & law' by Sartor (2009) and Sartor (2006)

[83] Directive 96/9/EC of the European Parliament and of the Council of 11 March 1996 on the legal protection of databases [1996] OJ L 77/20

[84] See De Beer (2016) and Rasmussen (2016)

[85] In such critical perspective, see Carbonell (2016) and Wolfert et al. (2017)

[86] As suggested, for instance, in Van der Burg et al. (2021)

agriculture such as plant protection products (PPP), that require approval before being placed in the market. Food business operators submitting data to the risk assessor (at EU level, the European Food Safety Authority (EFSA)) are not keen on releasing data into the public domain. A strong commercial interest in the protection of the investments needed to gather and analyse data can be easily justified for the research and development (R&D) costs incurred by the industry to place their new products on the market. Costs associated to the discovery, development and authorisation of a new plant biotechnology were calculated on a sum around $ 130m in 2011, spread across 15 years of investments from discovery to launch or more.[87]

However, these data also fall within the notion of environmental information for the purposes of EU law. As it will be observed in Chap. 2, this information is also subject to a certain degree of public interest. This is due to main reasons:

- On the one hand, scientific uncertainty polarises the debate on certain regulated products. As the effects of new foods, food materials or products on the medium-to-long term is unknown, tests have to be carried out to assess whether or not these products are safe. However, in the EU framework these studies are carried out by the industry (thus, with private funding) and transmitted to EFSA. Other entities (e.g., NGOs) want to access this information to cross-validate this scientific evidence.
- On the other hand, this information also concerns the environment and pieces of EU legislation allow the scrutiny of these data. When this is the case, the rationale underlying the disclosure mostly concerns the public nature of information related to the environment.

Either way, the need of data transparency is justified by the foreseeability of effects of certain products on human health and the environment. In practice, the balance is realised by a complex legal mechanism that provides, as a rule of thumb, that data transparency is the rule, whereas confidentiality is the exception.

Instead, the inverse relationship should the one for the personal data used during of risk assessment. The following section introduces the main topics related to the processing of personal data, whereas Chap. 3 will place this discussion within the current debate in data protection law.

1.2.2 Information Privacy, Food Consumption and Data Transmissions

In general, the processing of personal data can originate risks to rights and freedoms of natural persons (Recital 75 of the GDPR).[88]

[87] See the wide ranges proposed in McDougall (2011) and Bernauer et al. (2011)

[88] Regulation 2016/679 of the European Parliament and of the European Council of 27 April 2016 on the protection of natural persons with regard to the processing of personal data and on

In the EU, any processing activity carried out on personal data (i.e., any information relating to an identified or identifiable natural person[89]) is significantly regulated due to the relevance of personal data protection as a fundamental right (Art. 7 and 8 if the European Charter of Fundamental Rights), consistently with international law (Art. 8 of the European Convention of Human Rights).[90]

When contextualised to food-related information, the discussion pertaining the implications of personal data processing is convoluted by the need of balancing any restriction to data acquisition and analysis for privacy reasons with the necessary transparency of food safety risk assessment, the legitimate interests of food business operators and the exceptions in data protection law for research and statistical purposes. This balance—to which attention will be devoted in Chap. 3—seems necessary to ensure the free circulation of data among certain entities, whether they are involved in public tasks connected to food safety risk assessment, or pursue commercial purposes, while protecting individual rights and freedoms.

The literature has jointly discussed ownership rights of agriculture data also in terms of privacy,[91] thus highlighting concerns regarding information sharing practices and data security. However, this literature contains occasional misconceptions about the notion of "privacy", which often overlaps with "confidentiality". Such conceptualisation includes non-personal information, which falls outside the scope of data protection law[92] hence minimising risks to fundamental rights.[93] Instead, a more appropriate focus on personal data shall be the one pertaining to the datafication of food consumption, which occurs in several ways including food diaries and dietary surveys. This form of data processing, as it will be argued, can originate significant concerns related to information privacy of individuals whose data are processed.

Two levels of enquiry seem necessary to find sound trade-offs. On the one hand, the consequences of data acquisition and analysis shall be evaluated at the individual level of data subjects, i.e., the identified or identifiable individuals to whom data are

---

the free movement of such data, and repealing Directive 95/46/EC [2016] OJ L 119 (General Data Protection Regulation, GDPR). For a detailed commentary on the GDPR see the exhaustive discussion by Kuner et al. 2020) and compliance efforts are required for all the entities that collect, analyse, or share personal data. In particular, European institutions are subject to Regulation 2018/1725 (Regulation 2018/1725 of the European Parliament and of the Council of 23 October 201 8on the protection of natural persons with regard to the processing of personal data by the Union institutions, bodies, offices and agencies and on the free movement of such data, and repealing Regulation (EC) No 45/2001 and Decision No 1247/2002/EC [2018] OJ L 295/39)

[89] Article 4 of the GDPR states that personal data' means any information relating to an identified or identifiable natural person ('data subject'); an identifiable natural person is one who can be identified, directly or indirectly, in particular by reference to an identifier such as a name, an identification number, location data, an online identifier or to one or more factors specific to the physical, physiological, genetic, mental, economic, cultural or social identity of that natural person

[90] Lynskey (2014)

[91] Sykuta (2016)

[92] Leone (2017)

[93] Van der Burg et al. (2021)

referred to. On the other hand, implications of data processing can conveniently be considered at group level to include the social and collective dimension.

1.2.3 Big Data and Competition in the Food Industry

Finally, it is also worth mentioning that the opportunities given by the joint analysis of personal and non-personal information can be discussed under the perspective of the competitive advantage made possible by data-supported better decision-making. Market manipulation thanks to food-related information can occur both in cross-sectoral market sectors and within specific food-related fields.

The 2018 Bayer-Monsanto merger has created a *de facto* data conglomeration that makes possible the collection of experimental, agricultural and personal information, control over scientific publications and resources to compute large quantities of data by deploying cutting-edge methodologies. While it is still not possible to ascertain the informational implications of this merger, its consequences might range from the competitive advantage offered by such extraordinary amount of information to the possibility of commercialising high-quality data with high inferential capability.

As noted by the OECD in the same year,[94] Big Data can provide advantages for companies willing to develop their product, minimise costs, forecast market trends, segment and target customers (also in real-time), and so forth. The availability of large and structured quantities of data is transformed into valuable opportunities for market growth. This is related to certain properties of digital information, including its low—often null—reproduction costs. While control over Big Data does not necessarily entail a competitive advantage[95] due to factors other than data availability,[96] digital information could be exploited for anticompetitive purposes, including anti-competitive mergers, consumer discrimination, and collusion (also by algorithms). When discussing the implications of the Bayer-Monsanto merger from the perspective of competition policy, Commissioner Vestager stated that "[a]s digital agriculture and the use of Big Data become increasingly important, the question of who controls access to the data matters here, too".[97] While being heavily linked to *ownership-as-control*, competition issues related to Big Data seem also related to other factors.

[94] See the document 'Big Data: Bringing Competition Policy to the Digital Era' https://one.oecd.org/document/DAF/COMP(2016)14/en/pdf, accessed on 27/04/2022

[95] See Maggiolino (2018)

[96] See Lerner (2014)

[97] See 'Statement by Commissioner Vestager on Commission decision to give conditional approval to Bayer's plans to buy Monsanto and decision fining producers of capacitors €254 million for participating in a cartel', available at https://ec.europa.eu/commission/presscorner/detail/en/IP_18_2322, accessed on 27/04/2022

In its 2016 recommendations to antitrust authorities, the OECD identified certain criteria that shall be used to assess the status of the market prior to any intervention:[98] *inter alia*, (a) the degree of replicability of data in a given market,[99] (b) the substitutability between alternative datasets,[100] and (c) the quantity of necessary data to enter the market.

With specific regard to farming, scholars[101] have discussed the implications of restricting information sharing by *lock-in* mechanisms to the advantage of well-established players, confining the benefits of Big Data to the providers only for what relates solely to agri-business tasks *strictu sensu* (planting and harvesting) without considering non-agriculture information (e.g., crop price prediction), autonomy in participating Big Data platforms, and possible unfair competition among farmers that access to Big Data against those who cannot. Therefore, the nature of the information and the processing operations at stake have to be taken into account equally to control over and access to information.

1.3 Novel Trends in Food-Related Data Processing

This section is devoted to identifying some emerging trends in the processing of food related information. One challenge posed by this book is to identify a common approach to this subject matter: considering the wide range of implications of food safety risk assessment (public health, the environment, business interests, trust, and credibility of the institutions, etc.), it is necessary to underline some elements of novelty proposed by this book.

They encompass:

(a) an emerging approach to ethical issues pertaining to data, algorithms and practice ('data ethics'), and to governance ('soft ethics'), discussed in Sect. 1.3.1;
(b) a specific focus on food consumption data processing, whose implications are discussed in Sect. 1.3.2;
(c) an attempt to consolidate our discussion in the light of the forthcoming EU legislation on data governance and artificial intelligence (Sect. 1.3.3).

---

[98] See the document 'Big Data: Bringing Competition Policy to the Digital Era Executive Summary', available at https://one.oecd.org/document/DAF/COMP/M(2016)2/ANN4/FINAL/en/pdf, accessed on 27/04/2022

[99] The centrality of replicability of data mirrors the replicability of market strategy by competitors of a market player identified by Barney (1991). As noted, it might be the case that non-replicable data pave the way for non-replicable market strategies for competitors

[100] The role of substitutability in the approach of the EU Commission is further analysed by Cowen (2016) in relation to replicability and the epistemic value of Big Data

[101] See Sykuta (2016), Verdonk (2019), and Van der Burg et al. (2019)

1.3.1 'Data Ethics' and 'Soft Ethics'

Emerging food technologies are multi-faceted. Finding common traits among them is thus necessary to guarantee sufficient generalisability of the findings and consistency with the proposed holistic perspective towards the subject matter. Data ethics[102] may represent a viable way to transpose such convoluted technological framework into a systemic research scenario and provide solutions in a more speculative fashion. This is thanks to the duality of data ethics in being both descriptive and 'softly' prescriptive.

As regards the descriptiveness of data ethics, Floridi and Taddeo clarify the role of computer, information and data ethics by highlighting their different levels of abstractions (LoA).[103] LoA_I (the LoA of information ethics[104]) is mostly information-centric, the LoA_D (the LoA of data ethics) is clearly data-centric. While "the shift from information ethics to data ethics is probably more semantic than conceptual",[105] a specific focus on the moral dimension of data is undoubtedly convenient for the purpose of this book. However, as information ethics is one of the forerunners of data ethics, contribution in this domain will be taken into account[106]

LoA_I and LoA_D differ under some perspectives. LoA_I sets as observable the different moral dimensions of information when it is a source, the result or the target of moral actions. Hence, the modelling, sharing, retention, protection, use and deletion of information constitute possible variables of LoA_I. Data ethics and LoA_D are focused on three different sets of intertwined moral problems, namely those related to "data (including generation, recording, curation, processing, dissemination, sharing and use), algorithms (including Artificial Intelligence, artificial agents, Machine Learning and robots) and corresponding practices (including responsible innovation, programming, hacking and professional codes)".[107]

As regards the prescriptiveness of data ethics, its contributions are necessary every time a data-related ethical dilemma—i.e., choosing between two or more possibilities that, in principle, are equally right (or equally wrong)—emerges.[108]

---

[102] Floridi and Taddeo (2016)

[103] See Floridi (2008)

[104] As described in Floridi (2006)

[105] Floridi and Taddeo (2016)

[106] Instead, the perspective of Computer ethics in Moor (1985) another forerunner of data ethics, will not be discussed. The reason underlying this choice lies on the difficulty of framing LoA_C within the scope of this book, which is limited in terms of understanding the impact of computing in the society and in the environment, that is the goal of LoA_C

[107] Floridi and Taddeo (2016)

[108] This likely controversial statement shall be clarified. If data ethics is contextualised as one of the emergent fields of applied ethics, it follows that it inherits all the advantages and disadvantages of this category. In particular, in his review of critical approaches to applied ethics, Fossa (2017) noted that some philosophers struggle to recognise it as a serious philosophical discipline due to an alleged lack of rigorousness and the possible hybridisation of moral philosophy and professional practices. However, we agree with the Author when he also notes that "application, in ethics, has

As it will be noted, the emergence of ethical dilemmas is quite common in this domain. The task of data ethics is to maximise the "ethical value of data science to benefit our societies, all of us and our environments".

Data ethics should be read as a form of soft ethics. Following Floridi's interpretation of the role of ethics in this debate[109]—either as a challenge to the existing legislation to be used in a *de iure condendo* perspective ("hard ethics") or as what ought and ought not to be done *over and above* the existing regulations ("soft ethics"). The choice between "hard" or "soft" ethics consideration ultimately depends on multiple factors, including the degree of regulatory interventions in certain fields and the nature of the entity that commits to ethical behaviours.

With regards to voluntary commitments to ethics, Floridi identifies five possible threats, namely ethics shopping, ethics bluewashing,[110] ethics lobbying,[111] ethics dumping,[112] and ethics shirking.[113] while these concerns will be further explored in Sect. 5.3, it is necessary to bear in mind that 'soft ethics' can suffer from these limitations in practice.

The centrality of data in this book should not be confused with a quantitative methodology, nor should it be understood as limitation to static aspects of data processing. The state of the art in food-related data analysis suggests that confining the research domain to a static perspective would be wrong, as equal interests should be devoted to 'raw' information and to the way in which we 'make sense' of it. While certain issues—e.g., confidentiality and data protection—revolve mainly

---

primarily to do with the reasons that give moral meaning to our behaviours". Such behaviour, contextualised by the Author in terms of hermeneutic experience (in turn derived from Aristotle's practical philosophy), reflects morally-oriented solutions to problems occurring in practice. Hence, Fossa's hermeneutic theory of applied ethics is an attempt to reconcile moral philosophy and applied ethics.

[109] See, for an introduction to the concept, Floridi (2018)

[110] "Ethics Bluewashing" is defined as "the malpractice of making unsubstantiated or misleading claims about, or implementing superficial measures in favour of, the ethical values and benefits of digital processes, products, services, or other solutions in order to appear more digitally ethical than one is". Color blue is to differentiate Bluewashing from "Ethics Greenwahsing", that is the malpractice of appearing more environmental-friendly or sustainable than the actuality of things. See Floridi (2019)

[111] "Ethics Lobbying" is defined as "the malpractice of exploiting digital ethics to delay, revise, replace, or avoid good and necessary legislation (or its enforcement) about the design, development, and deployment of digital processes, products, services, or other solutions" See Floridi (2019)

[112] "Ethics Dumping" is defined as "the malpractice of (a) exporting research activities about digital processes, products, services, or other solutions, in other contexts or places (e.g. by European organisations outside the EU) in ways that would be ethically unacceptable in the context or place of origin and (b) importing the outcomes of such unethical research activities". See Floridi (2019)

[113] "Ethics Shrinking" is defined as "the malpractice of doing increasingly less ethical work (such as fulfilling duties, respecting rights, and honouring commitments) in a given context the lower the return of such ethical work in that context is mistakenly perceived to be". See Floridi (2019)

about raw data, the significance of algorithms and practices shall not be overlooked by the "data ethics" label.

1.3.2 The Untapped Relevance of Food Consumption Information

The aforementioned food consumption information represents a (qualitatively) significant segment of the data-types that are processed in the context of food safety risk assessment. While such relevance will be discussed in depth in the next chapters, it is first necessary to point out some limitations on the current debate about this information-type. Specific attention will be then devoted in Chap. 3.

In general, food consumption data can be described as records of foods eaten by individuals. Food consumption shall be actual or presumed. Hence, while this information differs from market preferences (e.g., food bought in a supermarket and recorded by a fidelity card), the former can be inferred from the latter (e.g., a supermarket chain assumes that the purchase of X quantity of food equals to its consumption in quantity Y). When this is the case, market preferences can qualify as food consumption information if labelled as such.

Dietary information can be recorded in several ways, both by third-party or via self-assessment. Among the first category of collection, telephone-based or computer-assisted methodologies are the most frequent recording methods for food safety purposes. Instead, food diaries constitute the most popular self-assessment method. Other, less relevant, examples are consumptions recorded in company canteens (e.g., associated to an employee) or schools.

When records of food eaten are associated to an identified or identifiable physical person, food consumption data qualify as personal for the purposes of data protection law. The association may sometimes be so weak that it is impossible to say who ate what (e.g., in food orders for families or other groups). In other instances, such as food diaries, the association is quite strong and actual or presumed consumption is inevitably associated to a single, identifiable individual.

As it will be discussed, food consumption data aggregation is extremely frequent in food safety because measurement of exposure to risks are performed at group level. Data aggregation usually amounts to a form of data anonymisation, even though this is not always the case as observed by data protection authorities,[114] including the European Data Protection Board (*rectius, the Article 29 Working Party*). Data aggregation is not exempt from issues related to the nature of inferences that such grouping allows. As it will be discussed, they can be used both to generate *regular* inferences, i.e., information categorised as "regular" personal data associated to specific individual, or *sensitive* inferences, i.e., information that qualifies as special categories of data for the purposes of EU data protection

[114] FTC (2015) and PDPC (2018)

law. Among regular inferences, market preferences are undoubtedly the most predominant category; among the sensitive ones, food consumption data can be used to infer the health status, philosophical and religious opinions, and (despite limitations) political positions.

Being a reliable proxy variable, i.e., an in-between information-type between inferable and inferred sensitive data, food consumption information has been traditionally studied in the context of other information-types whose processing seemed concerning from a data protection perspective. This is indicative of certain difficulties in finding clarity about the processing of special categories of information. In 2011, the Article 29 Working Party (A29WP) Advice Paper on special categories of data ("sensitive data") noted that 'due to the wide range of personal data that may fall into the category of health-related data, [. . .] this category represents one of the most complex areas of sensitive data and one where the Member States display a great deal of legal uncertainty'.

A comprehensive approach seems necessary to broaden the scope of the debate on this information-type. Such perspective could be beneficial to update this debate to novel pieces of legislation and processing techniques and to cast light on an overlooked field of research with peculiar traits including the highly inferential power and the necessity to balance access and distribution rights with the protection of individuals.

1.3.3 Big Data, AI, the Role of Principles in the Forthcoming Regulation

As mentioned at the beginning of this Chapter, this book proposes an "ethical Blueprint" in Chap. 5. I will refer to it as the P-SAFETY (Privacy, Security, Fairness, Explainability, Transparency) model, i.e. a list of principles aimed at promoting ethically-oriented behaviours in the food safety domain, inferred from established institutional ethical frameworks and ongoing regulatory works, interpreted in the light of the literature on ethics and law.

One may wonder whether such principle-based discourse is detached from the current debate on Big Data and AI governance. This subsection discusses how principles and, broadly speaking, ethics relate to this regulatory framework in the EU, in particular by considering forthcoming pieces of legislation including the AI Act[115] and the Data Governance Act.[116]

---

[115] Proposal for a Regulation of the European Parliament and of the Council laying down harmonised rules on Artificial Intelligence (Artificial Intelligence Act) and amending certain union legislative acts COM(2021) 206 Final

[116] European Commission Proposal for a Regulation of the European Parliament and of the Council on European data governance (Data Governance Act) COM/2020/767 final

The assumption underlying the idea that Big Data and AI should be governed by means of a principle-based regulation is consistent with the idea that these technologies should be used in a human-centric manner, i.e., fostering the human development. As Chap. 5 will note, this approach is somewhat inherently Kantian.

An ethics-oriented discussion has taken place in the EU before these proposals have been made, both at the EU level and within Member States. Chapter 5 will extensively report the main findings of this discussion. Here, it can preliminarily be noted that such principle-based pre-legislative discourse has taken place in at least three areas:

(i) Academic discussion. The contribution of disciplines such as the aforementioned data ethics, AI ethics, privacy and data protection, law & governance, legal informatics in this research area (that is, ethics of new technologies) has increased in the last years. Notably, this discussion has been enriched by contributions from data science and "pure" AI. The place of discussion has been academic conferences and journals;

(ii) Multi-stakeholder *fora*. Commercial actors, the academia and regulators have met in multi-stakeholder *fora* to discuss about principles that should steer the development of AI systems. Notable examples include the Asilomar Conference that will be discussed in Chap. 5.

(iii) Advisory boards. EU and Member States institutions have appointed expert boards to provide advice to policymakers. Chapter 5 discusses some examples. It might be too early to discuss their influence on the legislation, as the legislative process has not been yet finalised at the time of writing.

Reviewers have identified a significant agreement among these sources.[117] Due to such consensus, we can think that it might be possible to extend high-level ethical documents from these sources. They are not exempt from significant limitations.

As regards the involvement of the industry, scholars have refereed to this trend as "ethics washing"—i.e., a trend described as the use of "ethics" language intended to prevent legislative actions by imposing principle-based forms of self-regulation, while making AI practices palatable for the public.[118] To tackle this issue, In 2018 the European Commission declared that "[w]hile self-regulation can provide a first set of benchmarks against which emerging applications and outcomes can be assessed, public authorities must ensure that the regulatory frameworks for developing and using AI technologies are in line with these values and fundamental rights. The Commission will monitor developments and, if necessary, review existing legal frameworks to better adapt them to specific challenges, in particular to ensure the respect of the Union's basic values and fundamental rights.[119] This

---

[117] See Jobin et al. (2019), Morley et al. (2019), and Fjeld et al. (2020)

[118] Calo (2017), p. 408

[119] See p. 7 of the Communication from the Commission to the European Parliament, the European Council, the Council, the European Economic and Social Committee and the Committee of the Regions on Artificial Intelligence for Europe,Communication 'Artificial Intelligence for Europe'

justifies the institutionalisation of AI codes of conducts introduced by Title IX of the proposed AI Act and will be discussed in Chap. 5.

1.4 Chapter Synopsis

This introductory chapter has introduced the topic of this book by identifying *what* will be covered, *why* this range of topics is relevant in the current discussion on data ownership and data governance, and *how* this subject matter will be discussed. Food safety is a prime area of concern for this book. It represents a novel research field that is becoming popular thanks to innovation both in the technical and legal spheres. These two areas of discussion require a holistic approach capable of integrating findings emerging from these fields with the current debate on Big Data and AI governance. To this end, this book investigates how data ownership and governance should be regulated in the area of food safety, keeping into account principles inspiring the current and forthcoming legislation on emerging trends in Machine Learning and Big Data technologies.

Sub-topics of discussion cover the careful balance that must be struck between data confidentiality and transparency, both with regards to non-personal and personal information, algorithmic transparency, fairness in risk assessment, accountability frameworks, and so on. New legislative initiatives inside and outside our field of investigation cast new light over old debates.

The discussion on food safety technologies is not isolated. Food-related topics, such as Precision Agriculture or Big Data and competition, are already under discussion from the same "ownership & governance" perspectives. Food safety represents a natural progression of these studies. Moreover, AI law, governance, and ethics represent the benchmark used to validate our hypothesis.

This book aims to also be a reason of interest and a worthwhile goal for researches involved in the field of data and AI ethics, as it constitutes and example of applications of high-level principles to an applied and relatively unknown field. Such integration is realised by a principle-based proposal, the P-SAFETY model, that will be presented at the end of this book alongside some possible implementations which might constitute the integration of the model in real-life scenarios.

References

Accenture (2017) Future of food: new realities for the industry. https://www.accenture.com/us-en/_acnmedia/pdf-70/accenture-future-of-food-new-realities-for-the-industry.pdf. Accessed 24 April 2022

Accenture (2020) Digital agriculture: improving profitability. https://www.accenture.com/us-en/insights/interactive/agriculture-solutions. Accessed 24 April 2022

Alaimo LS, Fiore M, Galati A (2020) How the Covid-19 pandemic is changing online food shopping human behaviour in Italy. Sustainability 12(22):9594

Alemanno A (2014) Unpacking the principle of openness in EU law: transparency, participation and democracy. Eur Law Rev 1:72–90. ISSN 0307-5400

Alemanno A, Gabbi S (2016) Foundations of EU food law and policy: ten years of the European food safety authority. Routledge, Milton Park

Balafoutis AT, Beck B, Fountas S, Tsiropoulos Z, Vangeyte J, van der Wal T, Soto-Embodas I, Gómez-Barbero M, Marcus Pedersen S (2017) Smart farming technologies–description, taxonomy and economic impact. In: Precision agriculture: technology and economic perspectives. Springer, Berlin, pp 21–77

Barney J (1991) Firm resources and sustained competitive advantage. J Manag 17(1):99–120

Baumüller H (2017) Towards smart farming? Mobile technology trends and their potential for developing country agriculture. Handbook on ICT in developing countries. River Publishers, Delft, pp 191–201

BBC (2018) All protection steps taken' after BSE diagnosis. BBC news. https://www.bbc.com/news/uk-scotland-north-east-orkney-shetland-45954225

Bernauer T, Tribaldos T, Luginbühl C, Winzeler M (2011) Government regulation and public opposition create high additional costs for field trials with GM crops in Switzerland. Transgenic Res 20(6):1227–1234

Bevir M (2012) Governance: a very short introduction. Very short introductions. OUP Oxford, Oxford. ISBN: 978-0-19-164629-4. https://books.google.it/books?id=ozjcWIfhoO8C

Bielecka E (2020) GIS spatial analysis modeling for land use change. A bibliometric analysis of the intellectual base and trends. Geosciences 10(11):421

British Academy and Royal Society (2017) Data management and use: governance in the 21st century. https://royalsociety.org/-/media/policy/projects/data-governance/data-management-governance.pdf

Bronson K, Knezevic I (2016) Big data in food and agriculture. Big Data Soc 3(1):2053951716648174

Brown P, Will RG, Bradley R, Asher DM, Detwiler L (2001) Bovine spongiform encephalopathy and variant Creutzfeldt-Jakob disease: background, evolution, and current concerns. Emerg Infect Dis 7(1):6

Van der Burg S, Wiseman L, Krkeljas J (2021) Trust in farm data sharing: reflections on the EU code of conduct for agricultural data sharing. Ethics Inf Technol 23:185–198

Calo R (2017) Artificial intelligence policy: a primer and roadmap. UCDL Rev 51:399

Carbonell I (2016) The ethics of big data in big agriculture. Int Policy Rev 5(1). Original publication is online at https://policyreview.info/articles/analysis/ethics-big-data-big-agriculture

Carrefour (2018) Food blockchain. https://www.carrefour.com/en/group/food-transition/food-blockchain. Accessed 24 April 2022

Chang H-H, Meyerhoefer CD (2021) COVID-19 and the demand for online food shopping services: empirical evidence from Taiwan. Amer J Agricult Econ 103(2):448–465

Chen M, Mao S, Zhang Y, Leung VCM et al (2014) Big data: related technologies, challenges and future prospects, vol 96. Springer, Berlin

Corea F (2016) Big data analytics: a management perspective, vol 21. Springer, Berlin

Corea F (2019) AI knowledge map: how to classify AI technologies. In: An introduction to data. Springer, Berlin, pp 25–29

Cowen T (2016) Big data as a competition issue: should the EU commission's approach be more careful. Eur Netw L Reg Q 4:14

Cowls J, Tsamados A, Taddeo M, Floridi L (2021) A definition, benchmark and database of AI for social good initiatives. Nat Mach Intell 3(2):111–115

Curry E (2016) The big data value chain: definitions, concepts, and theoretical approaches. In: New horizons for a data-driven economy. Springer, Cham, 29–37

Dagar R, Som S, Kumar Khatri S (2018) Smart farming–IoT in agriculture. In: 2018 international conference on inventive research in computing applications (ICIRCA). IEEE, Piscataway, pp 1052–1056

Dannenberg P, Fuchs M, Riedler T, Wiedemann C (2020) Digital transition by COVID-19 pandemic? The German food online retail. Tijdschrift Voor Economische en Sociale Geografie 111(3):543–560

De Beer J (2016) Ownership of open data: governance options for agriculture and nutrition. Global Open Data for Agriculture and Nutrition, Wallingford [Report]. Available at SSRN: https://ssrn.com/abstract=3015958

De Mauro A, Greco M, Grimaldi M (2015) What is big data? A consensual definition and a review of key research topics. In: AIP conference proceedings, vol 1644, 1. AIP, pp 97–104

Durante M (2019) Potere computazionale: L?impatto delle ICT su diritto, società, sapere. Mimesis

EFSA (2011) Shiga toxin-producing E. coli (STEC) O104:H4 2011 outbreaks in Europe: taking stock. EFSA J 9(10):2390. https://doi.org/10.2903/j.efsa.2011.2390. https://efsa.onlinelibrary.wiley.com/doi/abs/10.2903/j.efsa.2011.2390

EFSA (2014) The European union summary report on trends and sources of zoonoses, zoonotic agents and food-borne outbreaks in 2012. EFSA J 12(2):3547

EPRS (2016) Precision agriculture and the future of farming in Europe: scientific foresight study. European Parliament Research Service, Brussels, Belgium

European Commission (2020) Commission White Paper on Artificial Intelligence - A European approach to excellence and trust. https://ec.europa.eu/info/sites/info/files/commission-white-paper-artificial-intelligence-feb2020_en.pdf. Accessed 26 April 2022

FAO (1983) World food security: a reappraisal of the concepts and approaches. Director general's report. FAO, Rome

FAO (2018) The state of food security and nutrition in the world 2018. http://www.fao.org/3/i9553en/i9553en.pdf

Faulkner A, Cebul K, McHenry G (2014) Agriculture gets smart: the rise of data and robotics. Cleantech agriculture report, Cleantech Group

Firican G (2018) Data governance maturity models - IBM | LightsOnData https://www.lightsondata.com/data-governance-maturity-models-ibm/

Fjeld J, Achten N, Hilligoss H, Nagy A, Srikumar M (2020) Principled artificial intelligence: mapping consensus in ethical and rights-based approaches to principles for AI. Berkman klein center research publication, nos. 2020–1

Floridi L (2006) Information ethics, its nature and scope. SIGCAS Comput Soc 36(3):21–36. ISSN: 0095-2737. https://doi.org/10.1145/1195716.1195719. http://doi.acm.org/10.1145/1195716.1195719

Floridi L (2008) The method of levels of abstraction. Minds Mach 18(3):303–329

Floridi L (2014) The fourth revolution: how the infosphere is reshaping human reality. OUP Oxford, Oxford

Floridi L (2018) Soft ethics and the governance of the digital. Philosop Technol 31(1):1–8

Floridi L (2019) Translating principles into practices of digital ethics: five risks of being unethical. Philosop Technol 32(2):185–193

Floridi L, Taddeo M (2016) What is data ethics? Philosophical Transactions of the Royal Society A: Mathematical, Physical and Engineering Sciences, 374(2083)

Fossa F (2017) What is moral application? Towards a philosophical theory of applied ethics. In: Applied ethics. The past, present and future of applied ethics, Center for Applied Ethics and Philosophy Hokkaido University (ed), pp 34–49

FTC, Federal Trade Commission (2015) Is aggregate data always private? https://www.ftc.gov/news-events/blogs/techftc/2012/05/aggregate-data-always-private. Accessed 22 April 2022

Gartner (2008) Gartner introduces the EIM maturity model. https://pdfs.semanticscholar.org/ca3b/13f65a37d7b0a44287899710112e2c5afc4e.pdf

Gavilan D, Balderas-Cejudo A, Fernández-Lores S, Martinez-Navarro G (2021) INNOVATION in online food delivery: learnings from COVID-19. Int J Gastron Food Sci 24:100330

Govindan K, Edwin Cheng TC, Mishra N, Shukla N (2018) Big data analytics and application for logistics and supply chain management. Transp Res Part E Log Transp Rev 114:343–349

Grimm-Samuel V (1991) On the mushroom that deified the emperor claudius. Class Quarterly 41(1):178–182

Grindle MS (2007) Good enough governance revisited. Develop Policy Rev 25(5):533–574
Hage, JC, von der Pfordten D (2009) Concepts in law, vol 88. Springer Science & Business Media, Cham
HLEG, AI (2019) A definition of AI: Main capabilities and scientific disciplines. %7B https://ec.europa.eu/digital-single-market/en/news/definition-artificial-intelligence-maincapabilities-and-scientific-disciplines%7D
Hoffman A, Munsterman R (2018) Dreyfus teams with banks for first agriculture blockchain trade. Available at https://www.bloomberg.com/news/articles/2018-01-22/dreyfus-teams-with-banks-for-first-agriculture-blockchain-trade. Accessed on 27/04/2022
Hohfeld WN (1917a) Fundamental legal conceptions as applied in judicial reasoning. Yale Law J 23(8):16–59
Hohfeld WN (1917b) Fundamental legal conceptions as applied in judicial reasoning. Yale Law J 26(8):710–770
IBM (2014) Infographic: the four V's of big data | IBM big data and analytics hub. https://www.ibmbigdatahub.com/infographic/four-vs-big-data
IBM (2016) The 5 V's of big data - Watson health perspectives https://www.ibm.com/blogs/watson-health/the-5-vs-of-big-data/
Jobin A, Ienca M, Vayena E (2019) The global landscape of AI ethics guidelines. Nat Mach Intell 1(9):389–399
Jordan MI, Mitchell TM (2015) Machine learning: trends, perspectives, and prospects. Science 349(6245):255–260
Kamath R (2018) Food traceability on blockchain: Walmart?s pork and mango pilots with IBM. J British Blockchain Assoc 1(1):3712
Kamilaris A, Fonts A, Prenafeta-Boldu' FX (2019) The rise of blockchain technology in agriculture and food supply chains. Trends Food Sci Technol 91:640–652
Knowles T, Moody R, McEachern MG (2007) European food scares and their impact on EU food policy. British Food J 1:43–67
Kuner C, Bygrave LA, Docksey C, Drechsler L (2020) The EU general data protection regulation (GDPR): a commentary. Oxford University Press, Oxford. ISBN: 978-0-19-882649-1. https://books.google.it/books?id=CBELtAEACAAJ
Ladley J (2012) Data governance: how to design, deploy and sustain an effective data governance program. The Morgan Kaufmann series on business intelligence. Elsevier Science, Amsterdam. ISBN: 978-0-12-397848-6. https://books.google.it/books?id=CpeAYWaTScYC
Laguna L, Fiszman S, Puerta P, Chaya C, Tárrega A (2020) The impact of COVID-19 lockdown on food priorities. Results from a preliminary study using social media and an online survey with Spanish consumers. Food Quality Pref 86:104028
Laney D (2001) 3D data management: controlling data volume, velocity, and variety. Technical Report. META Group. http://blogs.gartner.com/doug-laney/files/2012/01/ad949-3D-Data-Management-Controlling-Data-Volume-Velocity-and-Variety.pdf
Leone L (2017) Addressing big data in EU and US agriculture: a legal focus. Eur Food Feed L Rev 12:507
Lerner AV (2014) The role of'big data'in online platform competition. Available at SSRN 2482780
Logan D (2010) What is information governance? And why is it so hard? https://blogs.gartner.com/debra%7B%5C_%7Dlogan/2010/01/11/what-is-information-governance-and-whyis-it-so-hard/
Luckow A, Kennedy K, Manhardt F, Djerekarov E, Vorster B, Apon A (2015) Automotive big data: Applications, workloads and infrastructures. In: 2015 IEEE international conference on big data (big data). IEEE, Piscataway, pp 1201–1210
Lynskey O (2014) Deconstructing data protection: the added-value of a right to data protection in the EU legal order. Int'l Comp LQ 63:569
Lytos A, Lagkas T, Sarigiannidis P, Zervakis M, Livanos G (2020) Towards smart farming: systems, frameworks and exploitation of multiple sources. Comput Netw 172:107147
Maggiolino M (2018) I big data e il diritto antitrust. EGEA spa

Malgieri G (2016) Property and (intellectual) ownership of consumers' information: a new taxonomy for personal data. Privacy in Germany-PinG (4):133–150

Manyika J, Chui M, Brown B, Bughin J, Dobbs R, Roxburgh C, Hung Byers A, et al (2011) Big data: the next frontier for innovation, competition, and productivity. McKinsey Global Institute, Washington DC, USA

Maringer M, van't Veer P, Klepacz N, Verain MCD, Normann A, Ekman S, Timotijevic L, Raats MM, Geelen A (2018) User-documented food consumption data from publicly available apps: an analysis of opportunities and challenges for nutrition research. Nutr J 17(1):1–13

Mayer-Schönberger V, Cukier K (2013) Big data: a revolution that will transform how we live, work, and think. Houghton Mifflin Harcourt, Boston

McAfee A, Brynjolfsson E, Davenport TH, Patil DJ, Barton D (2012) Big data: the management revolution. Harvard Business Rev 90(10):60–68

McDougall P (2011) The cost and time involved in the discovery, development and authorisation of a new plant biotechnology derived trait. Consultancy Study for Crop Life International by P McDougall. Available at https://croplife.org/wp-content/uploads/pdf_files/Getting-a-Biotech-Crop-to-Market-Phillips-McDougall-Study.pdf. Accessed on 27/04/2022

Mendes J, Pinho TM, Neves dos Santos F, Sousa JJ, Peres E, Boaventura-Cunha J, Cunha M, Morais R (2020) Smartphone applications targeting precision agriculture practices? A systematic review. Agronomy 10(6):855

Miles C (2019) The combine will tell the truth: on precision agriculture and algorithmic rationality. Big Data Soc 6(1):2053951719849444

Mitchell TM et al (1997) Machine learning.McGraw-hill, New York

Mittelstadt B, Floridi L (2016) The ethics of big data: current and foreseeable issues in biomedical contexts. Sci Eng Ethics 22(2):303–341

Mittelstadt B, Allo P, Taddeo M, Wachter S, Floridi L (2016) The ethics of algorithms: mapping the debate. Big Data Soc 3(2):2053951716679679

Moody DL, Walsh P (1999) Measuring the value of information-an asset valuation approach. In: ECIS, pp 496–512

Moor JH (1985) What is computer ethics? Metaphilosophy 16(4):266–275

Morabito V (2015) Big data and analytics. Strategic and organisational impacts. Springer International Publishing, Berlin

Moréteau O, Masferrer A, Modéer KA (2019) Comparative legal history. Edward Elgar Publishing, Cheltenham

Morley J, Floridi L, Kinsey L, Elhalal A (2019) From what to how: an initial review of publicly available AI ethics tools, methods and research to translate principles into practices. Sci Eng Ethics 26:1–28

Morley J, Machado CCV, Burr C, Cowls J, Joshi I, Taddeo M, Floridi L (2020) The ethics of AI in health care: a mapping review. Soc Sci Med 260:113172

Mulla DJ (2013) Twenty five years of remote sensing in precision agriculture: key advances and remaining knowledge gaps. Biosyst Eng 114(4):358–371

Müller AC, Guido S (2016) Introduction to machine learning with Python: a guide for data scientists. O'Reilly Media, Sebastopol

Murnane EL, Huffaker D, Kossinets G (2015) Mobile health apps: adoption, adherence, and abandonment. In: Adjunct proceedings of the 2015 ACM international joint conference on pervasive and ubiquitous computing and proceedings of the 2015 ACM international symposium on wearable computers, pp 261–264

Oracle (2013) MDM maturity model. http://www.oracle.com/us/products/applications/master-data-management/mdm-maturity-model-1887940.pdf

Ostrom E, Calvert R, Eggertsson T (1990) Governing the commons: the evolution of institutions for collective action. Political economy of institutions and decisions. Cambridge University Press, Cambridge. ISBN: 978-0-52-140599-7. https://books.google.it/books?id=4xg6oUobMz4C

Pagallo U, Casanovas P, Madelin R (2019) The middle-out approach: assessing models of legal governance in data protection, artificial intelligence, and the web of data. Theory Pract Legislat 7:1–25

Pasquale F (2015) The black box society. Harvard University Press, Cambridge
PDPC, Personal Data Protction Commission Singapore (2018) Guide to basic data anonymisation technique. https://www.pdpc.gov.sg/-/media/Files/PDPC/PDF-Files/Other-Guides/Guide-to-Anonymisation_v1-(250118).pdf. Accessed 22 April 2022
Pierpaoli E, Carli G, Pignatti E, Canavari M (2013) Drivers of precision agriculture technologies adoption: a literature review. Procedia Technol 8:61–69
Poelman MP, Gillebaart M, Schlinkert C, Coosje Dijkstra S, Derksen E, Mensink F, Hermans RCJ, Aardening P, de Ridder D, de Vet E (2021) Eating behavior and food purchases during the COVID-19 lockdown: across-sectional study among adults in the Netherlands. Appetite 157:105002
Porter ME, Heppelmann JE (2014) How smart, connected products are transforming competition. Harvard Business Rev 92(11):64–88
Rasmussen N (2016) From precision agriculture to market manipulation: a new frontier in the legal community. Minn J L Sci Tech 17:489
Rossiter DG (1990) ALES: a framework for land evaluation using a microcomputer. Soil Use Manag 6(1):7–20
Russell S, Norvig P (2010) Artificial intelligence: a modern approach. Pearson Education, London
Samuel AL (1967) Some studies in machine learning using the game of checkers. II-Recent progress. IBM J Res Develop 11(6):601–617
Sartor G (2006) Fundamental legal concepts: a formal and teleological characterisation. Artif Intell Law 14(1):101–142
Sartor G (2009) Legal concepts as inferential nodes and ontological categories. Artif Intell Law 17(3):217–251
Seiner RS (2014) Non-invasive data governance: the path of least resistance and greatest success. Technics Publication. ISBN: 978-1-63-462045-1. https://books.google.it/books?id=Xo3XBgAAQBAJ
Singer JW (1982) The legal rights debate in analytical jurisprudence from Bentham to Hohfeld. Wis L Rev 975
Soares S (2015) Data governance tools: evaluation criteria, big data governance, and alignment with enterprise data management. Mc Press, Boise
Sonka S (2014) Big data and the ag sector: more than lots of numbers. Int Food Agribusiness Manag Rev 17(1030-2016-82967):1–20
Sykuta ME (2016) Big data in agriculture: property rights, privacy and competition in ag data services. Int Food Agribusiness Manag Rev 19(1030-2016-83141):57–74
Tamò-Larrieux A (2018) Mapping the privacy rationales. In: Designing for privacy and its legal framework. Springer, Berlin, pp 27–43
The Data Governance Institute (2015) Definitions of data governance - the data governance institute. http://www.datagovernance.com/adg%7B%5C_%7Ddata%7B%5C_%7Dgovernance%7B%5C_%7Ddefinition/
The Data Warehousing Institute (2013) Transforming data with intelligence. https://tdwi.org/pages/maturity-model/big-data-maturity-model-assessment-tool.aspx?m=1
Tian F (2016) An agri-food supply chain traceability system for China based on RFID & blockchain technology. In: 2016 13th international conference on service systems and service management (ICSSSM). IEEE, Piscataway, pp 1–6
Van der Burg S, Bogaardt M-J, Wolfert S (2019) Ethics of smart farming: current questions and directions for responsible innovation towards the future. NJASWageningen J Life Sci 90:100289
van Es HM , Woodard JD, Glos M, Verteramo Chiu L, Dutta T, Ristow A (2016) Digital agriculture in New York State: report and recommendations. Cornell University, Ithaca, NY
Verdonk T (2019) Planting the seeds of market power: digitalagriculture, Farmers'Autonomy, and the role of competition policy. In: Regulating new technologies in uncertain times. Springer, Berlin, pp 105–132

Vu Q, Raković M, Delic V, Ronzhin A (2018) Trends in development of UAV-UGV cooperation approaches in precision agriculture. In: International conference on interactive collaborative robotics. Springer, Berlin, pp 213–221

Wende K (2007) A model for data governance-organising accountabilities for data quality management. In: ACIS 2007 proceedings, p 80

Wolfert S, Ge L, Verdouw C, Bogaardt M-J (2017) Big data in smart farming-a review. Agricult Syst 153:69–80.

World Economic Forum (2020) This is how AI could feed the world?s hungry while sustaining the planet. https://www.weforum.org/agenda/2020/09/this-is-how-ai-could-feed-the-world-shungry-while-sustaining-the-planet/. Accessed 24 April 2022

Chapter 2
Data Ownership in Food-Related Information

2.1 Traditional and Emerging Approaches to Data Ownership

This chapter is devoted to review data ownership rules from a broad perspective encompassing 'black-letter law', jurisprudence of the European Court of Justice, and academic commentaries. The approach taken resembles the methodology of the legal formants proposed by Sacco[1] in the field of legal comparison, even though it is used for descriptive purposes.

Section 2.1 compares traditional ownership models with some emerging paradigms. This section deepens the general introduction to the topic provided in Sect. 1.2. Understanding existing ownership models is necessary to frame the evolution of the legislation and the relevance of the mentioned case law. Academic commentaries also provide insights on some emerging approaches that can contribute to a better conceptualisation of ownership in our domain, in particular when trust comes into play.

Section 2.2 explores the principles of Openness and Transparency in the current EU legal framework. In particular, it presents the Aarhus Convention and the Transparency Regulation, two cornerstones of the regulatory framework applicable to access to food-related data.

Section 2.3 critically evaluates the findings of the previous analysis and proposes a new concept of data ownership drawn from emerging models and their application to food safety information. Trust is proposed as the cornerstone of such novel approach.

[1] Sacco (1991)

S. Sapienza, *Big Data, Algorithms and Food Safety*, Law, Governance and Technology Series 52, https://doi.org/10.1007/978-3-031-09367-8_2

2.1.1 Ownership Models and Data

Section 1.2 has introduced the concept of data ownership and discussed how it will be used in this book. The concept has been defined as a bundle of control powers over data, including access, distribution, sharing, etc. It is now necessary to provide further justifications for this choice and discuss in detail its rationale and its background. Whenever something that is defined as 'data' is treated on a par with tangible objects, this information becomes a commodity, thus becoming 'good' on which legal rules may constitute rights, prescribe obligations, and so forth.[2]

The notion of property has been conventionally treated as a relationship between humans and tangible or intangible objects which constitute the subject matter of protection. In particular, intangible goods have fallen within the scope intellectual property law and several justifications—including human rights, utilitarian, and economical perspectives—have been proposed to as the rationale underlying their 'proprietarisation'.[3] However, the discussion about the relationship between humans and data have also been framed under perspectives others than property, in particular in the context of personal data. The straightforward applicability of property rights over data has been excluded on consequentialist reasons,[4] and none of the rationales provided to justify the existence of intellectual property seems straightforwardly applicable to personal data. Nonetheless, the debate has been framed in terms of a relationship between data subjects and individuals that can be expressed in the form of a scalar ownership[5] or controllership[6] to be evaluated according to a *case-by-case* analysis. However, a certain degree of ambiguity can be observed in the overlapping meanings of ownership and property, hence the necessity of re-conceptualise them for a clearer understanding in this book. This is also necessary to avoid inconsistencies with the scope of our investigation, which also includes other types of data.

The concept of ownership used for the purposes of this study is, in fact, reducible to a bundle of specific rights attributed by law, contracts[7] or other normative acts (e.g., a ruling) or facts (e.g., a newborn has—by default—rights over his or her personal data, even though they are managed by the holder of the parental responsibility). As argued in Sect. 1.2, the attribution of such rights originate corresponding (i.e. correlative or opposite) legal positions. The rightholder—original or derivative

[2] See also the introduction to the concept of data ownership and its relation on data governance in Sect. 1.2, in particular in the light of Hohfled's foundational work

[3] Aplin (2020), ch.1

[4] Samuelson (2000)

[5] British Academy and Royal Society (2019) and Malgieri (2016)

[6] Malgieri (2018)

[7] In particular, data ownership can be managed by means of licensing contracts. In this case, the original owner disposes of his or her property by attributing the totality or a part of his or her rights to another person

Table 2.1 Categorisation of goods in terms of excludability and rivalrousness

		Rivalrousness	
		Low	High
Excludability	Difficult	Public goods	Common-pool resources (CPRs)
	Easy	Toll or club goods	Private goods

following a legal act—can claim his or her right before competent authorities identified *ex ante* by the law or the legal act that attributes the right.

Ownership has been defined by the Oxford Dictionary of Law as 'an exclusive right to use, possess, and dispose of property, subject only to the rights of persons having a superior interest and to any restrictions on the owner's rights imposed by agreement with or by act of third parties or by operation of law'.[8] On the one hand, confusion arises since ownership pre-exists property, that is to say that ownership makes explicit the existence of a property-like legal relationship. On the other hand, ownership rights are mainly characterised by exclusivity, with the interference of external agents being exceptional. However, in a 'data ownership' scenario the concept of 'exclusivity' gains a significant importance due to the nature of the commodity at stake.

Consistently with this exclusion-based discourse, rules that prevent or allow external agents from accessing data can generate at least four models of ownership. Scholars who attempted to categorise commodities in terms of *exclusion* (i.e., how difficult is to exclude a person from enjoying a given good) have also included the concept of *substractability* or *rivalrousness* (i.e., whether the use made of an agent of a certain good prevents others from using the same good). Following this approach, they can be defined either as public, private, club goods, or as common-pool resources (CPRs) depending on their excludability and rivalrousness (Table 2.1).[9]

One recurrent misunderstanding is that data are not rivalrous *by-default*. It is a fact that data reproduction costs are infinitely lower than their material counterparty[10] and this factor has tricked many into thinking that rivalrousness of data cannot merely exist, as a simple duplication (read, 'copy-paste') which would allow any potential user to access digital information. However, this is not always true, since technological or legal constraints may be put in place to prevent access

[8] Martin (2009)

[9] Borgman (2015)

[10] Let us consider the reproduction cost of a book and its PDF/EPub version. While the costs associated to duplicate the physical book consist of printing, shipping, human resources, time, etc, its digital counterparty only necessitates a few click and, possibly, the electricity necessary to perform the computations linked to the reproduction. Since for most of digital items electricity costs are neglectable (this is not the case, for instance, of blockchain-based technology), reproduction costs of digital items are said to be zero

to data in a way that only a limited number of users at the time can be given access to certain information.[11]

Some data are characterised by low subtractibility/rivalrousness and difficult exclusion. This is the case of all the information which is made available to the public without any technological, economic or legal restrictions. For instance, 'open data'—broadly defined as those data 'in an open format that can be freely used, re-used and shared by anyone for any purpose'[12]—can fall under the category of public goods.

Club goods are characterised by easy excludability but low rivalrousness. In the 'material' world, a golf course is a club good since it offers a member-only entrance for a large, non-rivalrous space. In the context of this book, datasets that are subject to subscription-based access rules—such as case-law online repositories—or that can be accessed only under certain technical conditions—such as Application Programming Interfaces (APIs) that provide data in a proprietary and non-interoperable format—can be considered club goods.

The definition of CPRs refers to natural or man-made resources that are sufficiently large to make expensive—but not virtually impossible—to prevent others from obtaining benefits from its use.[13] A highway is a good example of CPR: it is hard to exclude others from riding it, but it might be subject to congestion. In the immaterial scenario, certain servers allow for a limited amount of possible simultaneous access to the dataset that they keep online.[14] This entails that, despite being (computationally) expensive, individuals could theoretically occupy the hundreds or thousands of spots available to access the data, for instance when several devices query the database.

Private data are characterised by easy excludability and high rivalrousness. In this sense, they constitute 'private property' of their owner. The legal attribution of exclusive rights over data may derive from intellectual property rights (IPRs).[15] As observed, 'personal data' are often treated just like 'private data' by scholars[16] supporting an ownership-based approach to data privacy (described

[11] Mahmoud (2019) and DECODE project (2020)

[12] Recital 16 of the Directive (EU) 2019/1024 of the European Parliament and of the Council of 20 June 2019 on open data and the re-use of public sector information [2019] OJ L 172/56). See also the definition by Global Open Data Index (2019), i.e., data that "anyone can freely access, use, modify, and share for any purpose (subject, at most, to requirements that preserve provenance and openness)'

[13] Ostrom et al. (1990), ch.2

[14] It is reported that data kept in a SQL server can be simultaneously accessed by 32.767 users Microsoft (2019)

[15] It worth noticing that, in EU law (Directive 96/9/EC of the European Parliament and of the Council of 11 March 1996 on the legal protection of databases [1996] OJ L 77), IPRs are related to databases. Copyright and *sui generis* protection are respectively intended to preserve the intellectual creation of the database structure and its content if a substantive investment in obtaining, verifying or presenting the database has been made

[16] I will discuss extensively privacy theories and their relevance for food consumption data processing in Chap. 3

Table 2.2 Categorisation of data in terms of excludability and rivalrousness

		Rivalrousness	
		Low	High
Excludability	Difficult	Open data	Data with limited access
	Easy	Subscription-based repositories	IPR-protected data

by Floridi[17]) and Restricted Access/Limited Control (RALC) theories.[18] The Restricted Access/Limited Control (RALC) and ownership-based theories interpret data privacy as the right to limit the amount of information available to other entities and are rooted on the psychological desire of secrecy and intimacy ('right to be let alone'). Following RALC-wise theories, the access to certain 'zones' or 'spheres' is restricted by individuals to protect themselves from external intrusion because personal data were subject to property rights.

Each presented scenario (Table 2.2) calls for an agent that exercises some degree of power over data: in private data, the agent is the person granted of exclusive rights by the law or by contract; in subscription-based repositories, the agent is the platform that makes the content available to the users; in 'open data' scenarios, control depends on the conditions under which data is made available to the public (e.g., Creative Commons licences), but it is safe to assume that we are confronted with a decentralised form of control democratically exercised through an entity (e.g., public administration); CPRs data can be owned by various entities (e.g., consortia), depending on the way in which the resource is managed.[19]

Data ownership can be then conceptualised as a bundle of rights (e.g., access, distribution, deleting[20] etc.) related to informational assets that show some degree of control over the information.

Since copies of the same data are perfect substitutes, the same informational asset can fall within multiple ownership models at the same time. This property becomes most relevant in particular when Machine Learning analysis generate value from the extracted information.[21] This property of digital information raises notable issues. Especially, moral dilemmas arise when deciding which ownership model shall be

[17] Floridi (2005)

[18] See the complete review by Tamò-Larrieux (2018)

[19] Ostrom et al. (1990), Ch. 3

[20] The inclusion of data erasure among the faculties attributed to a data owner is somewhat problematic. If *A* attributes some degree of ownership to *B* over data *XYZ*, then it might be the case that *B* can legitimately dispose of *XYZ* to erase them. It has been noted that more than one ownership models can be applied at the same time to the same data. Therefore, it is necessary to differentiate how this erasure can occur: *B* could either destroy his/her copy or erase *A*'s copy. The two cases are relatively common: the former case applies, for instance, when users erase copies of e-books in their e-shelves; the latter usually occurs under licensing conditions, e.g., when digital platforms prevent access to e-books copies due to subscription termination or non-payment.

[21] Duch-Brown et al. (2017)

adopted and why. To put it differently, who shall benefit from being the owner of information that can generate value?

2.1.2 Confidentiality and Digital Commons

It has been suggested that the shift from an ownership model to another of the *exact same* encoded digital information is easier than its material counterparty. It is now time to support this claim. Let us imagine two copies of a book, namely a book and its *e*-version in PDF/ePub format. While the book—like all the physical items—is normally characterised by high substractibility (usually, only one person at the time can read that copy), the same digital copy can easily be enjoyed by two or more individuals at the same time. Even though ways to prevent such contextual use (e.g., IPRs, licenses, Digital Rights Managements (DRM) tools) can be put in place, these are 'artificial constraints'[22] having regard to the nature of data, which is characterised by very low—and often null—reproduction costs as described above. Hence, while transforming the physical book into a CPR by donating it to a library is subject to certain conditions (e.g the approval of a librarian regarding the conditions of the book), the immaterial shift only requires the upload on a private shared folder reachable by the individuals designated by the owner (let us say, his or her friends) according to access limitations set by the service provider.[23] with consequences e.g., on copyright.[24] The reversed process is equally feasible: if someone accesses a publicly available copy of an e-book, they can easily store it in a private server and make it accessible only through a subscription fee if their licence allows so. Instead, finding a way to turn a book that belongs to a public library into private would amount to theft even when the book has fallen in the public domain.

The impact in terms of the excludability goes hand to hand with subtractability: while the physical book in the public library exists *only once* in its tangible manifestation, thus making excludability hard but not impossible (e.g., if the library that stores that *unique* copy of the book is closed on Sundays), the immaterial copy is accessible 24/7/365 and infinite duplicates can be created to make the book even more available,[25] to the extent that the original owner will not have excludability

---

[22] Rodotà (2015), p. 132

[23] For instance, 'basic' Dropbox users can achieve a maximum of 20 GB of bandwidth and 100,000 downloads per day. See https://help.dropbox.com/en-uk/files-folders/share/banned-links, accessed on 27/04/2022

[24] With regard to this relationship, see also Spedicato (2016) on the access to digital knowledge and copyright

[25] The description of this process has been purposively simplified. In fact, the authenticity of copied data can always be challenged. For instance, bytes might be lost during the upload/download/copy due to compression algorithms that manipulate the file without compromising the human perception. As a result, while a user could perceive data as a *perfect substitute* when accessing them, some units of information might have changed from the machine's perspective. In our description,

powers over the book, which should conveniently be considered a public good at that point.

However, none of these shifts prevents original owners from enjoying their own digital copy of the book, nor other individuals with the permission to read the copy stored in the shared folder, or to access the same copy mirrored on a public website, nor the subscription-based platform to offer the same book to its customers. This property of digital information contributes to the second challenge to classic ownership models: while material commodities might fall under one model at a time, the *same exact* immaterial resource can simultaneously fall under numerous models.[26]

To summarise, two major challenges have been raised when trying to adapt the traditional ownership models to a data-centric discourse. On the one hand, the shift of immaterial resources from one model to another is simpler than in their material counterparty; on the other hand, the simultaneous presence of multiple ownership models on the same resource can occur more frequently than in the material world. Therefore, a strict adherence to the classic ownership model cannot be pursued. Nonetheless, excludability and subtractability represent a pragmatic description of their hidden premise, i.e., having *control* over certain resources.

Choosing between alternative data ownership options implies the necessity of selecting the right controller of the data. This dilemma has originated two different and contrasting positions:

- Private companies are willing to protect their data from other the exploitation by other entities. Since it is relatively easy to access or duplicate data, commercial firms want to keep them as private as possible. The law covers this position by granting measures to protect digital information by keeping them confidential. In particular, databases can be protected by intellectual property rights, contracts or sector specific legislation.
- Given their easy reproducibility, some have advocates for granting to the general public the widest possible access to certain data, especially when a substantial public interest in accessing and verifying them is in place. When this is the case, this position has argued in favour of understanding data as digital commons, i.e., public, non-rivalrous, accessible digital information.

As the next section clarifies, food-related data are not exceptional to this dualism.

---

we will simply refer to the human understanding of data which is usually not compromised in the course of data transfer

[26] See the DECODE project (2020)

2.1.3 Transparency, Truthfulness, Trust

In the last decade, data related to food have been subject to extensive debates regarding the appropriate ownership regime in Europe. Claims in favour of the widest possible access to data have faced the resistance of private companies willing to keep their data confidential due to strong commercial interests.

Claims towards data transparency have revolved around the necessity of cross-validating scientific evidence gathered from data. While EFSA's opinions are published via their online journal as open-access articles, data remain subject to confidentiality rules, thus preventing a in-depth scrutiny of the way in which scientific evidence has been generated. The following two examples clarify why this mechanism has raised concerns among external entities willing to review this information.

In 2015, the International Agency for Research on Cancer (IARC) published a monograph about the carcinogenicity of the pesticide active substance glyphosate and claimed that chemical had met the criteria for classification as a 'probable human carcinogen'.[27] In the same year, EFSA replied to the Renewal Assessment Report (RAR)[28] request made by the German Federal Institute for Risk Assessment (Bundesinstitut für Risikobewertung, BfR) for glyphosate and concluded that the chemical had an unlikely carcinogenicity potential.[29]

Members of the European Parliament were denied access to confidential data, while some researchers claimed the existence of omissions (i.e., missing data) on the evidence submitted to the BfR.[30] As a reply, EFSA confirmed the adherence to standard protocols for data analysis and, therefore, the quality and the transparency of its review.[31] Following his letter, Portier and colleagues argued that 'to maintain transparency, IARC reviews only publicly available data. The use of confidential data submitted to the BfR makes it impossible for any scientist not associated with BfR to review this conclusion'.[32]

Then, let us consider the 'Monsanto Paper' scandal.[33] Dozens of pages of EFSA 2015 Renewal Assessment Report on Glyphosate were alleged to be copied by scientific papers 'ghostwritten' by Monsanto, a worldwide market leader in this segment.[34] Seemingly, PAN Europe NGO has raised doubts concerning the practices of a laboratory in Hamburg, alleged to 'distorting the data to please

[27] Fritschi et al. (2015)

[28] The RAR is a mandatory requirement for the continuation of the authorisation regime for pesticides. It should be submitted to a competent national authority and, in turn, to EFSA by the applicant of the renewal procedure

[29] EFSA (2015a)

[30] Portier (2015)

[31] EFSA (2015c)

[32] Portier et al. (2016)

[33] See, in the news, The Guardian (2017)

[34] McHenry (2018)

its clients'.[35] Collected data were also included in dossiers submitted during the evaluation procedure of pesticides and RAR.[36]

In these two scenarios, missing or altered data seem of particular concern since the possible mistakes in data collection might be replicated in further studies relying on such biased literature.[37] The amount of the risks linked to poor data quality escalates due to the increasing quantity of data and sources highlighted in the previous sections and the possible deployment of Machine Learning algorithms. When non-personal information is shared among the industry, independent laboratories, competent authorities at EU or Member States level, data quality has to be guaranteed. Since EFSA might not have sufficient resources to perform independent studies and its *ex novo* data generation is limited,[38] missing or altered third-party data represent a significant concern.

Data quality can be conceived as the epistemic dimension of truthfulness of data. Only high-quality information can guarantee the generation of reliable scientific evidence and, in turn, appropriate decisions. The low level of trust of EU food safety risk analysis mechanisms shown by the 2018 Fitness Check of the General Food Law[39] is a sub-optimal result also linked to the emergence of these scandals.

In the aftermath of the Monsanto Paper scandal, citizens' legislative initiative entitled "Ban Glyphosate and Protect People and the Environment from Toxic Pesticides"[40] was launched. Organisers collected 1,070,865 signatures. In summary, the petition contained:

(a) a proposal for a ban of glyphosate-based herbicides due to their carcinogenicity;
(b) a provision shifting the burden to prove the safety of regulated products for market approval from industry-led studies to institution-commissioned studies;
(c) the progressive reduction of pesticides use in the EU.

The Fitness Check covered a wide range of topics, including the perceived effectiveness of the principle of transparency.[41] The outcome was straightforward and it is worth quoting the Executive Report of the Fitness Check[42] in its entirety: '[d]espite overall considerable progress, transparency of risk analysis remains an

[35] PAN Europe (2020)

[36] At the time of writing, formal investigations are still running

[37] This issue becomes significant when in particular when Machine Learning techniques are deployed to explore the existing literature on a given topic. See Sect. 4.1.2

[38] EFSA (2015b), p.8

[39] EU Commission, 'REFIT Evaluation of the General Food law (Regulation (EC) No 178/2002' SWD(2018) 38 final, available at https://ec.europa.eu/food/sites/food/files/gfl_fitc_comm_staff_work_doc_2018_part1_en.pdf, accessed on 27/04/2022, hereafter 'Fitness Check'

[40] The official registration was recorded in the Communication of the Commission on the European Citizens' Initiative 'Ban glyphosate and protect people and the environment from toxic pesticides', available at https://ec.europa.eu/transparency/regdoc/rep/3/2017/EN/C-2017-8414-F1-EN-MAIN-PART-1.PDF, accessed on 27/04/2022

[41] Bartl (2015)

important issue in terms of perception: as regards risk assessment in the context of authorisation dossiers, EFSA is bound by strict confidentiality rules and by the legal requirement to primarily base its assessment on industry studies, laid down in the General Food Law Regulation and in the multiple authorisation procedures in specific EU food legislation. These elements lead civil society to perceive a certain lack of transparency and independence, having a negative impact on the acceptability of EFSA's scientific work by the general public. There is therefore a need to address these issues in order to protect the reputation of EFSA's work'. The full report clarifies how NGOs criticise that: (1) raw data are not made available due to confidentiality rules, thus making access to studies excessively restrictive, (2) secondary food legislation sets further confidentiality restriction, and (3) the combination of a general legal framework, secondary legislation and access to documents rules make 'create a rather complex system for the public release of documents'.[43]

In its reply to the 'Ban Glyphosate' initiative, the Commission stated its will to 'propose changes to the legislation to increase the transparency of studies commissioned by industry that are submitted in application dossiers, while respecting the principles [...] regarding the protection of legitimate confidential business information, including measures such as public access to raw data from study reports, thus reducing the need for stakeholders to have recourse to access to documents procedures'. This is exactly what happened with the Transparency Regulation.

2.2 Regulatory and Policy Trends in Openness and Transparency

Transparency measures applicable to public authorities can be broadly defined in two categories depending on the substantial role played by the public administration, i.e. reactive and proactive measures.[44] In reactive measures, the public body *reacts* to a request to access presented by an individual. In proactive measures, the administration *actively* makes documents and data available to citizens either voluntarily or to comply with legislative provisions that mandate the publication of certain information.

The rationales underlying transparency measures—further discussed below—range from the need of ensuring the democratic oversight of the working of public institutions, the release of data to foster the digital economy, the necessity to prevent corruption, historical or statistical purposes, allowing individuals to challenge the

[42] https://ec.europa.eu/food/system/files/2018-01/gfl_fitc_executive_summary_2018_en.pdf, accessed on 28/04/2022

[43] See the Fitness Check, p. 35

[44] Faini and Palmirani (2018)

decisions made by public bodies, and so on. Such wide span of justifications falls within the notion of 'open' government.

Among reactive measures, Regulation 1049/2001[45] is the cornerstone of EU's access to the data and the documents held by European institutions. It is grounded on the 'democratic oversight' rationale: Recital 2 states that 'Openness enables citizens to participate more closely in the decision-making process and guarantees that the administration enjoys greater legitimacy and is more effective and more accountable to the citizen in a democratic system. Openness contributes to strengthening the principles of democracy and respect for fundamental rights'. To this end, is guarantees the 'widest possible access' to data and documents held by the European Parliament, the Council and the Commission. Since 2021, EFSA is subject to the same rule due to the 2021 amendment to Art. 41 of the GFLR further discussed below.[46]

Exceptions to the general principle of transparency may apply and the institutions shall refuse the request to access data in the cases provided by Art. 2. *Inter alia*, the protection of the 'privacy and the integrity of the individual' (Art. 4(1)(b)), to be defined in accordance with data protection law and the protection of 'commercial interests of a natural or legal person, including intellectual property' (art. 4(2)) limit the remit of access to documents. Interestingly, an overriding public interest can still lead to the access of commercially sensitive data, whereas this derogation does not apply when personal data are at stake. Balancing exercises are then required to the institutions responsible to react to requests to access. As a rule of thumb, while transparency is the rule, data protection for privacy reasons or data confidentiality justified for commercial purposes are the exceptions.

As regards certain information, including food-related data, two further considerations shall be made:

- On the one hand, more specific reactive transparency measures are set by EU Regulations due to the nature of the information at stake. Such obligations further narrow the remit of exceptions to the benefit of transparency;
- On the other hand, reactive transparency measures are accompanied by proactive tools. While their rationales largely overlap, proactivity in this field is also grounded on the protection of public health in food safety matters due to the need of reviewing data and evidence on which decision-making processes are grounded.

---

[45] Regulation (EC) No 1049/2001 of the European Parliament and of the Council of 30 May 2001 regarding public access to European Parliament, Council and Commission documents [2001] OJ L145/43

[46] As noted in the Bavarian Lager case (T-194/04 Bavarian Lager v Commission [2007] ECR II-3201), the concept of 'document' for the purposes of Regulation 1049/2001 might include digital information held by the institutions subject to Regulation 1049/2001. Since the case concerned access to personal data contained in a minute (document), it is reasonable to assume that other forms of data (including non-personal) shall be subject to Regulation 1049/2001. A restrictive interpretation of the concept of documents seem inconsistent with other provisions on access to data and with the case law of the CJEU further analysed below

Conflicts may arise both when deciding the applicable laws and when balancing conflicting interests. This section first introduces the set of sector-specific reactive transparency measures relative to environmental information, then clarifies the scope of other provisions applicable to EFSA also in the light of the 2021 amendments. Then, it discusses some cases brought before the European Court of Justice.

2.2.1 The Aarhus Convention and the Right to Access Environmental Information

Regulation 1367/2006 ('Aarhus Regulation')[47] transposes the Aarhus Convention[48] within the EU legal framework and sets further obligations to Member States (who were already signatories of the Convention before the formal signature by the Union) to allow access to environmental information held by public authorities, including EU institutions.

The rationale underlying this regulatory framework is 'reserving, protecting and improving the quality of the environment and protecting human health, thereby promoting sustainable development' (Recital 1 of the Aarhus Regulation). Similar to the Convention, the Aarhus Regulation has a threefold aim: granting access to information related to emissions into environment, ensuring public participation in decision-making processes and providing access to justice in environmental matters. These pillars are correlated:[49] to contribute to decision-making processes in environmental decisions, citizens shall have access to information related to the environment held by public authorities and decision-makers, while disposing of redress mechanisms.

Art. 3 of the Aarhus Regulation extends the applicability of Regulation 1049/2001 to the environmental information held by EU authorities. 'Data or summaries of data derived from the monitoring of activities affecting, or likely to affect, the environment' are included among the list of environmental information that institutions shall keep organised in databases (art. 4(2)). Their formats shall be 'reproducible and accessible by computer telecommunications or by other electronic means' (art. 4(1)).

[47] Regulation (EC) No 1367/2006 of the European Parliament and of the Council of 6 September 2006 on the application of the provisions of the Aarhus Convention on Access to Information, Public Participation in Decision-making and Access to Justice in Environmental Matters to Community institutions and bodies [2006] OJ L 264/13

[48] Convention on Access to Information, Public Participation in Decision-Making and Access to Justice in Environmental Matters signed in Aarhus, Denmark, on 25 June 1998

[49] Wolf (2013)

The European Commission[50] has considered Aarhus Regulation to be largely convergent and coherent to Regulation 1049/2001. However, commentators[51] have noted that dual legislation can be hard to grasp for the citizens due to the interplay of several pieces of legislation.

As regards access to data, the Aarhus Regulation broadens the scope of the requests made to access under Regulation 1049/2001 to include environmental information specifically related to genetically modified organisms (Art. 2(d)(i)). Obligations coming from the Aarhus Regulation include the collection of 'environmental impact studies and risk assessments concerning environmental elements' (Art. 4(2)(g)). Then Art. 6(1) of the Aarhus Regulation prescribes a restrictive interpretation of the grounds of restriction of access set in Art. 4(2) of Regulation 1049/2001, which includes prejudices to 'commercial interests of a natural or legal person, including intellectual property'.

This is the first balance that is found by the legislator, i.e., a *de facto* rebuttable presumption mechanism of non-violation of confidentiality when access to environmental information is requested. This mechanism is justified by the necessity of promoting public health by reactively divulging documents that allow the scrutiny of independent entities. Notably, the presumption has not prevented the rise of conflicts in practice.

2.2.2 Transparency of Food Safety Information in the EU Following the 2021 Reform (Reg. 2019/1381)

2.2.2.1 What Information? The EFSA Data Warehouse

The EFSA Data Warehouse has been built to become the main data source for risk assessment activities performed at EFSA.[52] Data related to zoonotic diseases, antimicrobial resistance, foodborne outbreaks, pesticide residues, chemical contaminants, and chemical hazards collected yearly are accessible via this one-stop-shop hub. The database also contains data collected sporadically, i.e., *ad hoc* data collections derived from specific procurements.[53] The goal of this initiative is to strengthen scientific progress by granting access to data to food safety professionals, EFSA's stakeholders, and the general public.

[50] Commission Green Paper—Public Access to Documents held by Institutions of the European Community: A Review, COM 2007/0185 final

[51] Wolf (2013)

[52] EFSA (2011)

[53] EFSA (2015d)

Table 2.3 EFSA data warehouse access rules

Level of aggregation	Accessible data	Entitled users
Low	All data needed to perform their duties	EFSA staff member and EU Commission (DG SANTE)
	All data needed to perform their mandates	Members of EFSA Panels and Scientific Committee (limited to the time necessary to perform their mandates)
Medium	Data provided by their own organisation	Data providers
High	Data presented by EFSA's scientific and technical outputs	Data providers (following EFSA's output)
		EFSA's stakeholders (NGO's, academia, national food safety authorities, etc.)
		General public

EFSA Data warehouse is supported by access policies[54] which formalise the levels of access and granularity of accessible data in precise rules.[55] Access levels can be summarised as shown in the Table 2.3.

The granularity that defines the lowest level of aggregation is defined by EFSA in *ad hoc* agreements with data providers and the European Commission. Therefore, it depends on the information-type at stake and might differ according to the stakeholders involved (e.g., academia, food business operators, etc.). Since Member States qualify as data providers, their level of access is limited to the amount of data that they provide. However, pilot studies on Member States that agree to share their data at a more granular level by entering in 'Circles of Trust' have been conducted from 2014 to 2016.[56] The goal of these projects was to strengthen knowledge sharing among Member States as regards laboratory practices, analytical methods and monitoring programmes.

As regards data collection, the majority of data in the area of occurrence of chemical and biological hazards is submitted to EFSA by national food safety authorities in EU Member States.[57] Data providers also include the food industry, consumer associations and the European Commission. EFSA also publishes continuous 'calls for data'[58] to further fill and keep updated its Data Warehouse using voluntary contributions.

[54] Gilsenan (2015)

[55] EFSA (2015d)

[56] EFSA (2016)

[57] Kocharov (2009)

[58] https://www.efsa.europa.eu/en/calls/data, accessed on 26/04/2022

2.2.2.2 General Scope of the GFLR: Legal Basis, Goals and the Precautionary Principle

Today, the EU food safety legal framework consists of several sets of rules regulating diverse phenomena related to the food chain, from production to consumption and controls. This wide range of measures thus includes ingredient-related provisions (including rules on the use of additives, flavourings, enzymes) and consumer protecting norms (provisions on labelling, advertising, packaging, consumer rights). Harmonisation also covers administrative requirements and procedures, including the application for the placement of regulated products on the market. Given such wide range and the data-centric enquiry, the first focus will be on the general legal framework, which illustrates the global principles that frame sectoral legislation and EFSA's competences.

The General Food Law Regulation (GFLR)[59] is the cross-sectoral legislative framework in the EU food safety system. It has been drafted taking into account the 1997 Green Paper and the 2000 White Paper and sets the ground for the sectoral legislation covering specific food-related matters. Its legal bases are Articles 37, 95, 133 and Art. 152(4)(b) of the Treaty establishing the European Community,[60] which include objectives towards common agricultural policy, harmonisation, single market strategy and the promotion of public health. These goals are explicitly mentioned in Recital 1.[61]

Following the White Paper, risk[62] analysis is tripartite.[63] Together with risk management[64] and risk communication,[65] 'risk assessment' is defined as 'a scientifically

[59] Regulation (EC) No 178/2002 [2002] OJ L 31/1

[60] Consolidated Version of the Treaty establishing the European Community [2002] OJ C 325/33

[61] Recital 1 of the GFLR: [t]he free movement of safe and wholesome food is an essential aspect of the internal market and contributes significantly to the health and well-being of citizens, and to their social and economic interests

[62] Art. 3(9) of the GFLR: 'risk' means a function of the probability of an adverse health effect and the severity of that effect, consequential to a hazard

[63] Art. 3(10) of the GFLR: 'risk analysis' means a process consisting of three interconnected components: risk assessment, risk management and risk communication; Recital 7 of the GFLR: Where food law is aimed at the reduction, elimination or avoidance of a risk to health, the three interconnected components of risk analysis—risk assessment, risk management, and risk communication—provide a systematic methodology for the determination of effective, proportionate and targeted measures or other actions to protect health

[64] Art. 3(10) of the GFLR: 'risk management' means the process, distinct from risk assessment, of weighing policy alternatives in consultation with interested parties, considering risk assessment and other legitimate factors, and, if need be, selecting appropriate prevention and control options

[65] Art. 3(13): 'risk communication' means the interactive exchange of information and opinions throughout the risk analysis process as regards hazards and risks, risk-related factors and risk perceptions, among risk assessors, risk managers, consumers, feed and food businesses, the academic community and other interested parties, including the explanation of risk assessment findings and the basis of risk management decisions

based process consisting of four steps: hazard identification, hazard characterisation, exposure assessment and risk characterisation' (Art. 3(11) of the GFLR).

A notable provision regards the precautionary principle,[66] the precautionary principle is codified—for the first time in a normative statement[67]—in the Regulation. Art. 7(1) of the GFLR states that, 'in specific circumstances where, following an assessment of available information, the possibility of harmful effects on health is identified but scientific uncertainty persists, provisional risk management measures necessary to ensure the high level of health protection chosen in the Community may be adopted, pending further scientific information for a more comprehensive risk assessment'. Art. 7(2). specifies the need of these measures to be necessary and proportionate.[68]

While the precautionary principle is still a guiding principle for risk managers, it illustrates the necessary interplay between risk assessment and risk management with the latter being directly influenced by the quality ('nature of the risk to life or health') and quantity ('the type of scientific information needed') of scientific uncertainty. As risk managers *may* take measures on the basis of the precautionary principle, such interplay cannot be deemed as a necessary dependence. If we include the principle of proportionality and the temporal constrains to adopt measures, we can notice that a reasonable margin of discretion is still granted to risk managers in spite of the bindingness of the precautionary principle.

2.2.2.3 EFSA, Data Collection and Storage

Let us now turn to the establishment of EFSA and its task. Chapter III of the GFLR is dedicated to its mission and its internal organisation. Art. 22 attributes risk assessment competences (Paragraph 2) and requires their independent fulfilment (Paragraph 7). EFSA's tasks are listed in Art. 23. In addition to the drafting of the 'best possible scientific opinions' (Art. 23(a)) and support activities to risk management, they include the duty to 'search for, collect, collate, analyse and summarise scientific and technical data in the fields within its mission' (Art. 23(c)).

[66] Section 4.2.3 discusses in detail the rationale of the precautionary principle and its relevance in data-driven risk assessment

[67] Rusconi (2016)

[68] Art. 7(2) of the GFLR: [m]easures adopted on the basis of Paragraph 1 shall be proportionate and no more restrictive of trade than is required to achieve the high level of health protection chosen in the Community, regard being had to technical and economic feasibility and other factors regarded as legitimate in the matter under consideration. The measures shall be reviewed within a reasonable period of time, depending on the nature of the risk to life or health identified and the type of scientific information needed to clarify the scientific uncertainty and to conduct a more comprehensive risk assessment

Then, Art. 33(1) details the information-types that EFSA has to collect:

- food consumption and the exposure of individuals to risks related to the consumption of food;
- incidence and prevalence of biological risk;
- contaminants in food and feed;
- residues.

Paragraphs 2 and 3 of Art. 23 set further provisions regarding the collaboration between EFSA and national authorities in Member States, who are required to transmit the data listed above to EFSA, including those submitted by business organisations, academic studies and third countries. Finally, Paragraph 6 mandates that the European Parliament, the Commission and Member States shall be allowed to scrutinise EFSA's work in the field of data collection. Interestingly, 'the Authority shall search for, collect, collate, analyse and summarise relevant scientific and technical data' (Paragraph 1), whereas the scrutiny of risk managers and Member States is only extended to 'the results of its work in the field of data collection' (Paragraph 6). The reasons underlying this discrepancy are unclear. Perhaps, the legislator intended to broaden the scope of the 'field of data collection' mentioned above to include all the sets of operation listed in Paragraph 1, rather than confining the remit of assessors' oversight to the mere acquisition of data.

2.2.2.4 EFSA Data Transparency Obligations

The provisions set out in Art. 38, 39 and 41 aim at striking a balance between the clashing interests identified in the previous Chapter. Art. 38(1) lists the kind of document that EFSA must make public without delay. Together with internal agenda and minutes (lett. a), scientific opinions adopted by the Scientific Committee and the Scientific Panels (lett. b), declarations of interests (lett. d), annual reports (lett. e), and amendments to requests to scientific opinions from the European Parliament, the Commission or Member States (lett. f), two data-related kinds of documents have to be published for transparency reason:

- The information on which its opinions are based, without prejudice to Articles 39 and 41 discussed below (lett. (c)). The identification of this information is *prima facie* unclear. Several interpretations of the notion of information—ranging them from the literature on a given topic to the personal notes of scientists involved in the studies—might be proposed. In particular, should raw data be included among the information on which EFSA's opinions are based? Two answers, equally right in principle, should be investigated. On the one hand, data are epistemically functional to generate scientific evidence and hence a positive answer regarding the possible inclusion should be given; on the other hand, a negative answer might depend on a narrow interpretation of the term 'information' which is

inextricably linked to the level of abstraction of an epistemic agent[69] (Sect. 1.1.3), thus preventing the inclusion of raw data for being devoid of any semantic value.
- The results of its scientific studies (lett. (e)); similar to the previous point, questions about the information to be disclosed can be raised. In particular, shall derivative and inferred data be considered 'results' of EFSA scientific studies for the purposes of publication? Some might advocate that the remit of 'results' should be confined to the epistemic conclusions reached by EFSA after its studies, whereas others could argue that data analysis generates outputs that are different from scientific evidence nonetheless to be published for their instrumentality to a transparent and accountable risk assessment.

2.2.2.5 EFSA Confidentiality Management

Leaving aside temporarily these questions, it is time to focus on the remit of confidentiality as defined by Art. 39 of the GFLR. The general principle is enshrined in Paragraph 1: '[b]y way of derogation of Art. 38, the Authority shall not divulge to third parties' confidential information that it receives for which confidential treatment has been requested and justified, except for information which must be made public if circumstances so require, in order to protect public health'. A literal interpretation of the provision suggests that the secrecy of confidential information that does not belong to the Authority is (1) a derogation to the general principle of transparency and (2) subordinate to the protection of consumers' health.

EFSA is competent for the evaluation on confidentiality claims, decided according to an internal procedure that assesses the circumstances of individual claims brought by companies (Art. 39(1)). This internal procedure for the handling of confidentiality claims is regulated within the general Guidance for the application for regulated products.[70] The factors to be taken into account are: whether the information claimed to be confidential is available only to a limited number of individuals, and is not publicly available (*secrecy of the information*); whether the disclosure of the information at stake will result in serious harm to the interests of the person who has provided it or to third parties (*harmfulness of the disclosure*); whether the interests claimed to be harmed by the disclosure of the information at stake are worthy of protection *(worthiness of the interests*). The cited reference guide includes data (globally referred to as 'information' or 'scientific information') among the documents to be submitted.

Art. 41(1) states that Regulation 1049/2001 shall apply to documents held by the Authority, which in turn is obliged to adopt by reactive transparency measures (Art. 31(2)). In addition to these implementations, in 2015 EFSA decided to proactively open a portion of its Data Warehouse to the public,[71] in accordance with the access

[69] Durante (2019) and Floridi (2013)

[70] EFSA (2018), para 2.15

[71] EFSA (2015d)

rules discussed in Sect. 2.2.1. In 2019, EFSA has strengthened proactive disclosure of data via Zenodo—Knowledge Junction.[72] This platform is also used to identify metadata that contribute to data standardisation, as with the case of dietary and background information metadata.

In summary, the GFLR lays down several rules concerning the collection and dissemination of personal and non-personal data, to be read within the context of risk analysis (i.e., assessment, management and communication) and in light of the precautionary principle (i.e., uncertainty of risks entails that cautious measures have to be taken) and communication duties, including in the case of scientific uncertainty. This legislative framework has also been drafted under the umbrella of the concept of 'transparency', which is also the heading of Art. 38 of the GFLR. This principle has multiple facets, which include fostering trust in the institution (Recital 9 of the GFLR), contributing to risk assessment via scientific studies (Art. 9 of the GFLR), fostering a democratic oversight over decision-making processes.

The 2002 version of the GFLR is not exempt from a high margin of discretion in assessing the secrecy of information, the harmfulness of data publication and the worthiness of the interests underlying the non-disclosure of data, namely the three criteria that lead to a given decision on confidentiality. This might be due to the necessity of preserving flexibility when managing confidentiality requests or preventing conflicts thanks to a progressive dialogue between food business operators applying to authorisations and the Authority. However, flexibility makes it hard to foresee the outcome of the confidentiality claim (which is detrimental for the industry) and the negotiation, in the absence of a public *ex ante* scrutiny, leaves citizens with little or no oversight on the whole process. The exercise of discretionary powers by EFSA has also led to claims brought before the ECJ, whose outcomes are discussed later on in this Chapter. Instead, the next subsections will discuss the 2019 amendments to the GFLR that directly tackles these issues.

2.2.2.6 Regulation 2019/1381

The Transparency Regulation, i.e., Regulation 2019/1381[73] consists of a set of amendments to the GFLR and sectoral legislation following the Citizen's Legislative Initiative 'Ban Glyphosate' discussed in Sect. 2.1.3. The reform heavily amends transparency provisions and the data submission procedure. The first proposal from the Commission was published on 11 April 2018 and approved on 20 June 2019. Provisions entered into full effect on 27 March 2021.

[72] Available at https://zenodo.org/communities/efsa-kj/?page=1, accessed on 27/04/2022

[73] Regulation (EU) 2019/1381 of the European Parliament and of the Council of 20 June 2019 on the transparency and sustainability of the EU risk assessment in the food chain and amending Regulations (EC) No 178/2002, (EC) No 1829/2003, (EC) No 1831/2003, (EC) No 2065/2003, (EC) No 1935/2004, (EC) No 1331/2008, (EC) No 1107/2009, (EU) 2015/2283 and Directive 2001/18/EC [2019] OJ L231

Table 2.4 GFLR and transparency regulation

GFLR art.	Transparency regulation art.	Heading
32	32 (not replaced)	Scientific studies
	32a	Pre-submission Advice
	32b	Notification of studies
	32c	Consultation of Third Parties
	32d	Verification Studies
33	33	Collection of data
34–37 (not replaced)		
38	38 (replaced)	Transparency
39	39 (replaced)	Confidentiality
	39a	Confidentiality request
	39b	Decision on confidentiality
	39c	Review of confidentiality
	39d	Obligations with regard to confidentiality
	39e	Protection of personal data
	39f	Standard data formats
	39g	Information systems

Table 2.4 shows a direct comparison between the pre-reform version of the GFLR and the amendments brought by the Transparency Regulation.

2.2.2.7 New Articles 32a 'Pre-submission Advice', 32b 'Notification of Studies', 32c 'Consultation of Third Parties', 32d 'Verification Studies'

New articles 32a, 32b, 32c, 32d have been inserted. In the context of applications for regulated products, the Authority is now competent for giving advice on 'the rules applicable to, and the contents required for, the application or notification, prior to its submission' (New Art. 32a(1)), including via its website (New Art. 32a(2)). Art. 32b(1) establishes a database of studies commissioned or carried out by business operators prior to the application. Undertakings shall notify the Authority of the scope of the study, without delay (New Art. 32b(2)). Vice versa, laboratories shall notify EFSA the mandate to carry out studies (New Art. 32b(3)). The notification of the study is a mandatory requirement within the application procedure, even though remission mechanisms allow the re-submission (New Articles 32b(4), 32b(5)). New Art. 32c provides for a consultation mechanism in the case of a renewal of an application/authorisation.[74] When this is the case, EFSA has to launch a

[74] Certain regulated products (e.g., GM food and feed) are subject to a mandatory renewal procedure after a certain number of years (e.g., 10 years for Genetically Modified food (Art 7(5) of Regulation 1829/2003)

'consultation of stakeholders and the public on the intended studies for renewal, including on the proposed design of studies' following the notification of the studies (New Art. 32c(1)). Access to data by these third parties is granted only to the non-confidential version of the application dossier as described below (New Art. 32c(1)). Consultation has to occur immediately after the disclosure of the dossier 'in order to identify whether other relevant scientific data or studies are available on the subject matter concerned'. New Art. 32d allows the Commission to authorise EFSA to commission independent studies in exceptional circumstances of serious controversies or conflicting results. This provision might be deemed as the acceptance of petitioners' claims for more independence from the industry by EFSA. However, as these independent studies are limited to cases of scientific uncertainty—which might also emerge from *ex post* studies rather that prior research as observed in IARC case—and subject to budget constraints, the efficacy of Art. 32d can be challenged.

2.2.2.8 Amended Art. 38 'Transparency'

Art. 38(1) has been significantly amended by several new provisions. For what concerns data 'proactive' publication, subheadings specify the scientific information-types to be published:

- (b) EFSA scientific outputs, including the opinions of the Scientific Committee and the Scientific Panels after adoption, minority opinions and the results of consultations performed during the risk assessment;
- (c) Scientific data, studies and other information supporting applications, including supplementary information supplied by applicants, as well as other scientific data and information supporting requests from the European Parliament, the Commission and the Member States for a scientific output, including a scientific opinion;
- (d) Information on which EFSA scientific outputs, including scientific opinions, are based;
- (f) EFSA scientific studies in accordance with Art. 32 and new Art. 32d

The first three provisions mentioned above include the clause 'taking into account the protection of confidential information and the protection of personal data in accordance with Articles 39 to 39e'.[75] Information shall be made public without delay, except scientific data under letter (c) which shall be made public without delay once an application has been considered valid or admissible. Scientific outputs under letter (b) have to be made public in a dedicated section of the Authority's website open and accessible to the public. They should also be downloadable, printable and indexed.

[75] Its implications and mechanisms will be discussed below

With particular regard to data disclosure under letters (c) and (d), new provisions in the new Paragraph 1a of Art. 38 include the protection of:

- (a) existing rules concerning intellectual property rights which set out limitations on certain uses of the disclosed documents or their content;
- (b) any provisions set out in Union law protecting the investment made by innovators in gathering the information and data supporting relevant applications for authorisations ("data exclusivity rules")

Finally, Art. 38.1a sets out that the disclosure to the public of information 'shall not be considered to be explicit or implicit permission or licence for the relevant data and information and their content to be used, reproduced, or otherwise exploited in breach of any intellectual property right or data exclusivity rules, and the Union shall not be responsible for its use by third parties. The Authority shall ensure that clear undertakings or signed statements are given to that effect by those who access the relevant information prior to its disclosure'.

2.2.2.9 Amended Art. 39 'Confidentiality'

The amended version of Art. 39 opens with a reaffirmation of the exceptional nature of confidential treatment over submitted information (Paragraph 1). New Paragraph 2 imposes on applicants the burden to prove that the disclosure of the information at stake can potentially harm their interests. Moreover, it clarifies the information-types for which confidential treatment can be requested. Together with information on applicant's manufacturing processes and methods, commercial relationships and business strategy, lett. (d) extends the scope of confidentiality requests to the quantitative composition of the subject matter of the request, except for information which is relevant to the assessment of safety. Paragraph 3, allows derogations of confidentiality rules to be implemented by EU food sectoral legislation. Paragraph 4, derogates to the list of Paragraph 2 in the event of urgent actions to protect animal health or the environment or if the information at stake is the basis of a scientific output, including opinions, issued by the Authority when effects on the same subject are foreseeable.

2.2.2.10 New Articles 39a 'Confidentiality Request', 39b 'Decision on Confidentiality', 39c 'Review of Confidentiality', Art. 39d 'Obligations with Regard to Confidentiality'

Articles inserted after Art. 39 clarify the mechanism of the confidentiality request. Most notably, new Art. 39a provides that applicants requesting confidential treatment have to submit a non-confidential version of the dossier devoid of information deemed worth of confidentiality and elicited as missing. The full and confidential version shall also be submitted, with allegedly confidential information clearly marked.

New Art. 39b sets out the general procedure to follow when the Authority handles confidentiality requests, as the figure below displays in a simplified and accessible form. Together with the examination of the confidentiality requests, the publication of the non-confidential version of the dossier occurs without delay following the submission. The dialogue between applicants and EFSA is now regulated according to a precise timeline (Fig. 2.1).

Additional data can also be submitted during the procedure. However, this does not hamper their publication: as a rule of thumb, all the data for which confidentiality requests or claims have been rejected have to be made public, either without delay or following a 'cooling-down period' given to the applicant to reason about its next steps. Remarkably, reasoned decisions taken by EFSA can be now challenged before the Court of Justice (New Art. 39b(3)).

Art. 39c states that EFSA has to periodically review confidentiality decisions whether conclusions of scientific outputs, including scientific opinions, highlight foreseeable effects on human and animal health or on the environment and the previously confidential information has to be made public.[76] When this is the case, Art. 39b applies.

Art. 39d consists of a series of additional provision regarding confidentiality, e.g., for what concerns staff and managers or in the case of withdrawal of application. In relation to the procedure set out in Art. 39b, new Art. 39d(2) clarifies that information on which confidentiality request has been submitted shall not be made available to the public until a final decision has been reached. Likewise, the Commission and the Member States shall make efforts to prevent the publication of information for which confidentiality has been accepted.

On March 2020, EFSA published its practical arrangements on Articles 38 and 39.[77] Article 10 lists the substantive requirements of confidentiality requests, i.e., a clear indications of documents, information or data for which confidentiality requests can be made for which applicants shall provide explanation/justification to obtain the confidential status. In particular, applicants should prove:

- The secrecy of the information, to be intended as the lack of public availability.
- A potential significant harm of the disclosure, equivalent to the 5% of the total turnover for legal persons or earnings for natural persons. If this requirement is not met, a documented justification of a potential or foreseeable harm to applicant's interests has to be submitted;
- The worthiness of protection for the concerned data, i.e., the eligibility of data and information to be subject to legal protection;
- That data and information do not fall within the notion of "environmental information" pursuant to Article 2 of the Aarhus Regulation;

---

[76] This new provision mirrors 'reactive' transparency measures mandated by Regulation 1049/2001 as amended by the Aarhus Regulation. These measures will be discussed in Sect. 3.3

[77] Available at https://www.efsa.europa.eu/sites/default/files/corporate_publications/files/210111-PAs-transparency-and-confidentiality.pdf, accessed on 27/04/2022

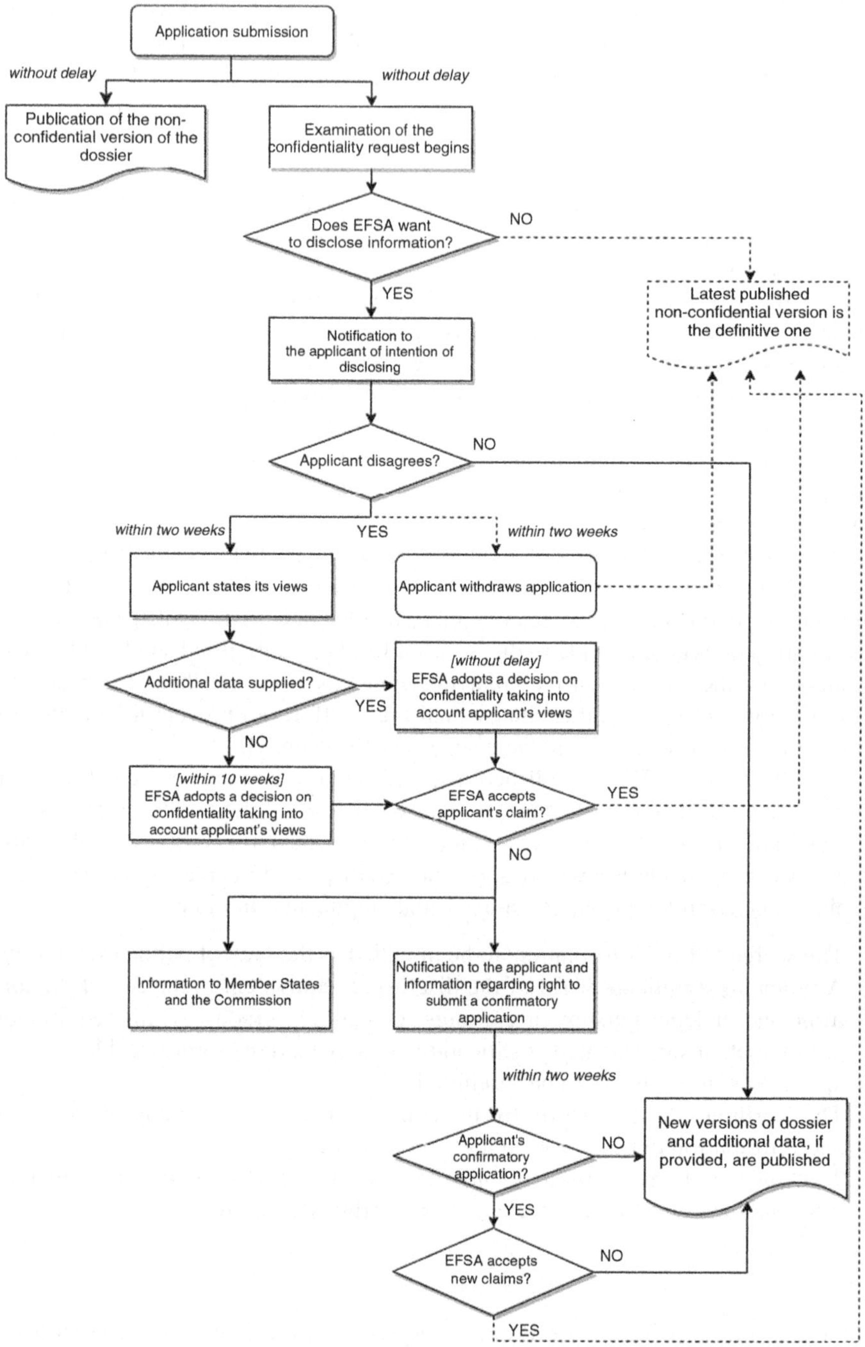

Fig. 2.1 Confidentiality requests management process (New Art. 39b GFLR)

- The confirmation that document, information or data, have been finalised within 5 years prior to the submission. If this requirement is not met, a specific justification of the potential harm to applicant's interests has to be attached.

The European Crop Protection Association (ECPA)[78] published its legal analysis of the draft practical arrangements.[79] ECPA noted that (1) EFSA might have set a threshold (5%) which cannot contemplate the loss in the competitive position of the applicant and the competitive advantage to competitors since they are too difficult to calculate prior to the disclosure of data, and (2) EFSA might have exceeded the scope of the Transparency Regulation when setting the 'not older that 5 year' requirement to assess the novelty of document, information and data. This latter requirement also allegedly lacks of a clearly identified rationale that links the 5-year timespan to the presumption of a harmless disclosure.

2.2.2.11 New Art. 39e: 'Protection of Personal Data'

Art. 39e regards the protection of personal data. However, the information that the legislator mainly intended to protect is of a different kind than the food consumption data discussed in Chap. 3. In fact, Art. 39e regulates the processing of four different kinds of information:

- information pertaining the name and address of the applicant, author(s) of supporting studies, participants and observers of working groups and scientific panels. For transparency reasons, these names have to always be made public.
- names and addresses of natural persons involved in testing on vertebrate animals or in obtaining toxicological information. The disclosure of this information is deemed to significantly harm the privacy and the integrity of these individuals, so it should not be made public unless differently specified by Regulation 2018/1725,[80] which balances the protection of personal data processed by EU bodies and institutions with the general access to the same information by the public for transparency reasons.[81]

[78] The European Crop Protection Association (ECPA) represents crop protection industry in the EU. It is registered in the EU Transparency Register (ID: 0711626572-26) among the trade and business associations, It promotes the 'the scientific study and analysis of all fields of interest to the crop protection industry'

[79] Available at https://ec.europa.eu/info/law/better-regulation/have-your-say/initiatives/12140-Food-safety-pesticide-approvals-clearer-information-/F555972_en, accessed on 27/04/2022

[80] Regulation 2018/1725 of the European Parliament and of the Council of 23 October 2018 on the protection of natural persons with regard to the processing of personal data by the Union institutions, bodies, offices and agencies and on the free movement of such data, and repealing Regulation (EC) No 45/2001 and Decision No 1247/2002/EC [2018] OJ L 295/39

[81] The reason underlying this provision is that '[u]nfortunately, a minority of individuals who oppose the use of animals in research have used threats and even terrorism to further their views' against researchers' (Beversdorf et al. (2015))

- any personal data made public pursuant to Art. 38 GFLR.[82] The divulgation of this data shall only be used to ensure the transparency of the risk assessment. A clear reference to the necessary compatibility of the secondary processing is provided by an explicit mention and Art. 4(1)(b) of Regulation 2018/1725.
- any other personal data the processing of which is carried out pursuant the GFLR (residual clause). This processing of this information by EFSA falls under the regime of Regulation 2018/1725.

The new provision on personal data seems to be mainly focused on the names of individuals working within the remit of risk assessment activities or other scientists involved in testing. However, we will note in Chap. 3 that EFSA also collects personal food consumption data and background information at individual level via Member States and makes them available though its Food Consumption Database in aggregated forms. The legal regime for the processing of this information is the one enshrined by the last point mentioned above, which covers *any* personal data processed in the context of the GFLR. It might be questioned whether or not this regime also applies to commercial applicants.[83] Following a literal interpretation of Art. 39e, an affirmative answer should be given insofar the statement 'the processing of personal data carried out pursuant' the GFLR is interpreted broadly. However, this would imply the extension of Regulation 2018/1725 way beyond its scope (Art. 2 of Regulation 2018/1725),[84] thus a cautious answer should be preferred, making the GDPR alone applicable to commercial applicants that operate within the material and territorial scope of the GDPR should be preferred.

2.2.2.12 New Art. 39f 'Standard Data Formats', Art. 39g 'Information Systems'

Transparency Regulation also intervenes on standardisation of data formats. It is required that the Authority and the Commission cooperate to draw formats that are not based on proprietary standards, are capable of ensuring interoperability with the state-of-the-art in research, and suits the need of small and medium-sized enterprises. The adoption of standardised data formats is then detailed in sectoral legislation and EFSA's practical arrangements.

Finally, new Art. 39g prescribes the use of auditable and secure information systems by EFSA, specifically to protect personal and confidential data.

---

[82] Art. 38 contains few explicit examples of personal data: participant lists (Art. 38(1)(a)) and declarations of interest (Art. 38(1)(e)) made by the members of EFSA working groups and scientific panels, authors of the scientific opinions (Art. 38(1)(b))

[83] It could be assumed that applicants for regulated products authorisation usually qualify as data controllers, whereas external laboratories to whom analysis is delegated should be deemed as data processors operating on behalf of the applicant under a data processing agreement

[84] Art. 2(1) of Regulation 2018/1725: [t]his Regulation applies to the processing of personal data by all Union institutions and bodies

2.2.3 *Recent Case Law of the European Court of Justice: A Short Commentary*

Another significant contribution to the debate surrounding transparency measures *vis-a-vis* confidentiality and data protection is the one made by the jurisprudence of the Court of Justice of the European Union (CJEU). Over the last years, its decisions have progressively adopted a 'more transparency' approach by confining the remit of confidentiality measures due to an overriding public interest in the disclosure of data. This section aims to reconstruct the judicial reasoning of the Court, the goals of transparency in its words, and how the balance of public and private interests has been struck in practice. Consistently with the scope of this study, our focus will be on cases that explicitly confront with data-related issues in the food safety risk assessment domain.[85]

This section refers to a specific sectoral legislation, i.e., the legal framework for the application to place Plant Protection Products (PPPs, usually referred to as 'pesticides') in the EU market. The applicable legislation consists of Regulation 1107/2009,[86] Commission Regulation 546/2011[87] and Commission Regulation 284/2013[88] which mandate the submission of data (Art. 8 of Regulation 1107/2009) by the applicant and a review by EFSA (Art. 12 of Regulation 1107/2009) following

[85] Nevertheless, some other cases might be of interest for contextualising the ones discussed in the next sections. *Inter alia*, Case T-70/99 *Alpharma Inc. v Council of the European Union* ECLI:EU:T:2002:210 [2002] ECR-II-03495 frames the scope of risk assessment and its relationship with risk management in light of the precautionary principle ('Notwithstanding the existing scientific uncertainty, the scientific risk assessment must enable the competent public authority to ascertain, on the basis of the best available scientific data and the most recent results of international research, whether matters have gone beyond the level of risk that it deems acceptable for society') [para 175] (Rusconi 2016, p.457). An other case settled on the same day—Case T-13/99 *Pfizer Animal Health SA/NV v Council of the European Union* EU:T:2002:209 [2002] ECR-II-03305—clarified that 'scientific advice on matters relating to consumer health must, in the interests of consumers and industry, be based on the principles of excellence, independence and transparency'. This wording is particularly significant due to the concomitant institution of EFSA, which was established a few months before the General Court's decision. The wording of the General Court in this case has been interpreted as a remark to provide rigorous standards in the definition of prerequisites of decision-making (Peel 2012). A similar approach has been taken in a later case—C-79/09 *Gowan Comércio Internacional e Serviços Lda v. Ministero della Salute* EU:C:2010:803 [2010] ECR I-13533 annotated by Alemanno 2011

[86] Regulation 1107/2009 of the European Parliament and of the Council of 21 October 2009 concerning the placing of plant protection products on the market and repealing Council Directives 79/117/EEC and 91/414/EEC [2009] OJ L 309/2009

[87] Commission Regulation 546/2011 of 10 June 2011 implementing Regulation (EC) No 1107/2009 of the European Parliament and of the Council as regards uniform principles for evaluation and authorisation of plant protection products [2011] OJ L 155/127

[88] Commission Regulation 284/2013 of 1 March 2013 setting out the data requirements for plant protection products, in accordance with Regulation (EC) No 1107/2009 of the European Parliament and of the Council concerning the placing of plant protection products on the market [2013] OJ L 93/85

a preliminary assessment of the Member State to whom the application was first referred. The same applies in the case of renewal. Transparency Regulation has harmonised the handling of confidentiality claims regarding PPPs to the procedure set out in the GFLR.

This legal framework is further enriched by the applicability of the Aarhus Regulation, which requires that EU bodies to make environmental information available upon request based on Regulation 1049/2001 and following an assessment of possible derogations while bearing in mind their restrictive interpretation. As discussed above, exceptions may include the protection of legitimate commercial interests of a third party, including intellectual property. However, Aarhus Regulation states the prevalence of the public interest in the disclosure.

As it will be shown, together with conclusions applicable to each controversy, this analysis identifies an interpretative mechanism put in place by the CJEU courts, i.e., a principle-based approach that is used to identify the rationale of reactive and proactive measures of data publication or access. Once clarified their scopes, these principles serve to identify criteria used to settle the dispute, usually by declaring the prevalence of the public interest in the disclosure. This leaves open questions regarding the possibility to extend the information to be made available (reactively or proactively) to other informational components (e.g., computer programmes and operational data) following the same approach and the same criteria adopted by the CJEU.

2.2.3.1 *Greenpeace and PAN Europe* on the Concept of Environmental Information and on the Balance of Public and Commercial Interests

In 2013, a 'remarkable' case[89] discussed the reactive disclosure of documents by the Commission when environmental information is at stake. In the controversy—*Greenpeace Nederland and Pesticide Action Network Europe v Commission*[90]—the General Court had to confront with the interpretation of what constitutes information relating to emissions into the environment. Defendants requested access to documents and data on the composition of a glyphosate-based pesticide [paras 2, 17]. Although EFSA was not directly involved in these cases, this controversy is worth of mention due to the discussion regarding the public disclosure of commercial information that originated from this case.[91] It is also significant under two profiles: on the one hand, the decision clarifies the scope of 'information *related* (emphasis added) to emissions into the environment' for the purposes of Aarhus Regulation, which applies to subject matters under the competences of EFSA; on the other hand, it points out whether the balance between public interest in the

[89] Moules (2017)

[90] Case T-545/11 *Stichting Greenpeace Nederland and PAN Europe v Commission* EU:T:2013:523

[91] Korkea-Aho and Leino (2017), p.1080

disclosure and the commercial interests in protecting valuable information should be done in abstract or by looking at concrete factors.

Applicants of the case requested access to environmental information on certain PPPs held by the Commission and transmitted by the German Government. The Commission partially rejected the request by claiming that over a specific volume, containing a 'complete list of all tests', the German authority had refused to provide access and the Commission was in turn obliged to act in compliance with its decision. German authorities had previously declared that they did not consider to be an overriding public interest for the disclosure of the concerned document[92] and the information contained were not pertaining to emissions into the environment [para 7]. The Commission opted to protect intellectual property rights over the public interest in the disclosure of data, deeming the previous publications adequate to ensure transparency [paras 9–11].

In the judgement, the General Court first restated that one of the goals of Regulation 1049/2001 is to favour open decisions to strengthen democracy [para. 27], consistently with its previous case-law. In *Turco*[93] the ECJ had observed, when referring to the goals of Regulation 1049/2001, that: '[o]penness in that respect contributes to strengthening democracy by allowing citizens to scrutinize all the information which has formed the basis of a legislative act. The possibility for citizens to find out the considerations underpinning legislative action is a precondition for the effective exercise of their democratic rights' [paras 41–43, 59–61].

Second, when confronting with the exceptions to the disclosure for reasons of protecting private interests, the General Court observed that 'since they derogate from the principle of the widest possible public access to documents, those exceptions must be interpreted and applied strictly' [para 32] and in line with TRIPS (Trade Related Aspects of Intellectual Property Rights) Agreement, when possible.[94] Therefore, the Court concluded that exceptions to the general rule of disclosure, including the intellectual property, must be interpreted and applied strictly to not to frustrate the goals of Regulation 1049/2001 [para 50].

Third, the General Court remarked that Recital 15 of the Aarhus Regulation calls for a restrictive interpretation of the grounds of refusal of public access. When

---

[92] In particular, it has been argued that, by giving access to the documents at stake, other competitors would have been able to to copy the production processes, thus leaving commercial interests and intellectual property rights unprotected. See Bazylińska-Nagler (2017)

[93] Case C-39/05 *Sweden and Turco v Council* [2008] ECR-I-04723

[94] While TRIPS agreement is outside the scope of this thesis, it has to be remarked that the Court also discussed the relationship between EU provisions and the TRIPS agreement, which is part of the internal legal framework of the Union. It stated that '[w]here there are European Union rules in a sphere concerned by the TRIPS Agreement, European Union law will apply, which will mean that it is necessary, as far as possible, to adopt an interpretation in keeping with the TRIPS Agreement, although no direct effect may be given to the provision of that agreement at issue' [paras 45–46]. For an extensive discussion on TRIPS and environmental information in this judgement, see the case note by von Holleben (2013), sec. II(2)

assessing the request, authorities shall consider two factors, namely the 'public interest the disclosure' and the relationship between the information and emissions into the environment [paras 52–53]. As regards the former, it is unclear whether the public interest should be evaluated *in abstracto* or *in concreto* terms.[95] Regarding the latter, the Court interpreted such relationship as a 'sufficiently direct link' [paras 53, 57], in a manner that the concept of 'information relating to emissions into the environment' also includes the contents of the dossier at stake [paras 60–61, 66]. On this ground, the General Court annulled the decision at issue by the Commission, insofar as it refused access to those parts of the dossier including information on 'emissions into the environment' as interpreted above. One commentator[96] has criticised the adoption of such criterion by claiming its extensive broadness, up to the point to run 'the risk of classifying the entire use of chemicals as emission'.

In the appeal,[97] the ECJ accepted the premise of interpreting exceptions to the divulgation of information related to emissions into the environment restrictively [para 50]. However, while adopting the same principle-based approach of the first instance, it rejected the 'sufficiently direct link' criterion and adopted a new standard, namely that the concept of 'emissions into the environment' covers 'information on current and foreseeable emissions', whereas it does not include 'purely hypothetical emissions'[98] [para 73–75].[99]

The teleological interpretation of the dissemination obligation is particularly significant. The ECJ stated that the concept access to environmental information 'must be understood to include, *inter alia*, data that will allow the public to know what is actually released into the environment', including foreseeable effects [para 79]. The list of information to be made available—'information concerning the nature, composition, quantity, date and place of the actual or current and foreseeable emissions, under such conditions, from that product or substance'—that follows this statement is a consequence of this rationale. Moreover, such dissemination is needed a) to include citizens in decision-making processes, the accountability of

[95] For instance, whether or not the commercial/non-governmental nature or the purposes of the requester should be kept into account or the Commission should evaluate them broadly. One commentator—von Holleben (2013), p. 571—has noted that the General Court found an irrebuttable statutory presumption in the conjunction between Art. 4(2) of Regulation 1049/2001 and the Aarhus Regulation that prevents any further balancing of conflict of rights by making the public interest in the disclosure prevailing *iuris et de iure* without the need of assessing the peculiarities of the request

[96] von Holleben (2013), p. 575

[97] Case C-673/13 P *Commission v Stichting Greenpeace Nederland and PAN* Europe [2016] ECLI:EU:C:2016:889. Noteworthily, the case saw written observations by CropLife International, National Association of Manufacturers of the United States of America, America Chemistry Council Inc., European Crop Care Association, European Chemical Industry Council, and Association européenne pour la protection des cultures in support of the Commission

[98] Moules (2017)

[99] One commentator (Bazylińska-Nagler 2017) has noted that other formulae ('information which relates to emissions' and 'information with a sufficiently direct link to emissions') contributed to the overall vagueness of the terminology used by the ECJ

decision-makers and the public awareness of environmental matters [para 80], and b) for reasons of consistency with the general aims of Regulation 1049/2001, i.e., 'the balance which the EU legislature intended to maintain between the objective of transparency and the protection of those interests' [para 81]. On this basis, the ECJ set aside the judgment of the General Court and referred it back.

As noted by a commentator,[100] the line between 'foreseeable' and 'purely hypothetical' emissions may sound unclear. On the same day, the Court delivered another ruling—*Bayer CropScience*[101]—in which it clarified that information on 'current and foreseeable emissions' consists of 'data concerning the medium to long-term consequences' of emissions and studies on the measurement of the substance's drift when realistic conditions—yet, 'most unfavourable' [para 91]—but it does not cover simulations in which normal or realistic conditions are not met due to 'significantly' higher doses [para 90].

Finally, the case came to its conclusion in 2018.[102] The General Court accepted the 'information on current and foreseeable emissions' criterion [para 57-58] and concluded that the Commission 'did not commit an error of assessment in considering that the document at issue does not contain information relating to emissions into the environment' [para 91], contrary to what stated in the previous judgement.

Moreover, the General Court also provided answers for the question pertaining whether the evaluation of the balance of public and commercial interests should be carried out in abstract or practical terms. The General Court opted for the latter approach by stating that 'it must be shown that the documents at issue contain elements which may, if disclosed, seriously undermine the commercial interests of a legal person' [para 110]. In the case at stake, the General Court did not confirm its previous decision and approved the balance of interests carried out by the Commission [paras 111–112].

In summary, the controversy provided answers for some interpretative doubts regarding the nature of 'environmental information' in the case of PPPs and how the balance between public and private interests regarding this information should be assessed. As regards the first issue, the information-types at stake are not relevant for our discussion. Instead, the criterion to assess whether the information should be made available is of particular interest. Two criteria, namely the 'sufficiently direct link' and 'information on current and foreseeable emissions' were adopted, with the latter prevailing in the end. Commentators from ClientEarth[103] noted that the 'foreseeable emission' criterion would hinder the possibility to access studies where higher doses are tested to verify acute and long-term exposure and proposed a

---

[100] Moules (2017)

[101] Case C-442/14 *Bayer CropScience SA-NV and Stichting De Bijenstichting v College voor de toelating van gewasbeschermingsmiddelen en biociden* [2016] ECLI:EU:C:2016:890

[102] Case T-545/11 RENV *Stichting Greenpeace Nederland and PAN Europe v Commission* ECLI:EU:T:2018:817

[103] Buonsante and Friel (2017), p. 457

broader interpretation that shall include all the documents used to draw conclusions about current and foreseeable emissions.

2.2.3.2 *Tweedale* and *Hautala* on the Overriding Public Interest on the Disclosure of Confidential Dossiers

On 7 March 2019, the General Court delivered two cases, *Tweedale*[104] and *Hautala*,[105] which directly concerned EFSA and its decisions.

Tweedale case pertains to the access to documents related to Glyphosate. In 2014, Anthony Tweedale, a consultant on environmental matters, requested EFSA to access documents related to two 'key studies used in order to set Glyphosate's acceptable daily intake (ADI)' [para 9], pursuant to Regulation 1049/2001. EFSA had these studies in its possession pending Glyphosate assessment. The Authority refused the access due to the exception related the protection of commercial interests (Art. 4(2) of Regulation 1049/2001).

According to EFSA, such protection was due to the confidential status that was granted in the application procedure [para 8]. EFSA also considered that the proactive publication of scientific information relating to the safety of Glyphosate was 'manifestly and fully satisfied' by the data made available on its website [para 12]. In 2017, following several rounds of discussion, EFSA decided to give access to raw data and findings of the two 'key studies' while also claiming that confidentiality status did not apply to that information [para 18–20]. However, the Authority gave only limited access and kept confidential some administrative and manufacturing information, as well as information related to the protocols followed by the study owners [para 21]. EFSA acted on the basis of a 'balance of interests' analysis and concluded that no overriding public interest could have been found in the non-disclosed information. It concluded that 'raw data and findings were sufficient to examine carefully the evaluation of the results of the requested studies' [para 23]. The case stayed pending until the final judgements of *Commission v PAN Europe* and *Bayer CropScience* [para 27] and resumed after their final delivery [para 31].

The applicant raised a total of six pleas, which could be summarised in two broad categories: on the one hand, that EFSA had misclassified the non-disclosed information in both decisions (first and fifth claims); on the other hand, that EFSA had let the private interest in keeping data confidential prevail over the public one in disclosure while failing to provide sufficient justifications to keep the undisclosed data away from public eyes (second, third, fourth, and sixth claim).

[104] Case T-716/14, *Anthony C Tweedale v European Food Safety Authority* [2019] ECLI:EU:T:2019:141

[105] Case T-329/17, *Heidi Hautala and Others v European Food Safety Authority* [2019] ECLI:EU:T:2019:142

Hautala case has a similar factual background. In 2016, Heidi Hautala and other four MEPs requested access to glyphosate-related scientific studies held by EFSA, this time by justifying their request by highlighting the IARC- EFSA controversy. As with *Tweedale*, EFSA granted access to a large *corpus* of unpublished studies, including raw data and findings, while keeping material, experimental conditions and methods secret, likewise results and discussions.[106] EFSA relied on a twofold rationale: on the one hand, the disclosure would have harmed the commercial interest of the concerned company [paras 22–23]; on the other hand, the requested studies did not fall within the remit of 'environmental information' to be reactively disclosed [para 24]. Pleas made by applicants were substantially the same as the ones in *Tweedale*.[107]

As regards the scrutiny of the balance of public and private interests, the Court consolidated its narrow interpretation of the confidentiality exceptions with respect to general principle of transparency. The Court remarked that, in Regulation 1049/2001, 'openness enables the EU institutions to have greater legitimacy and to be more effective and more accountable to EU citizens in a democratic system and that, by allowing divergences between various points of view to be openly debated, it also contributes to increasing those citizens' confidence in those institutions' [*Tweedale*, para 75; *Hautala*, para 60]. Such teleological interpretation of the rationale underlying Regulation 1049/2001 is not entirely new since the Court had already introduced it in *ClientEarth v Commission*.[108]

With regards to the nature of approval dossiers of PPPs submitted by applicants for authorisation, they were qualified as 'environmental information' by the Court. The rationale was the same used in *Stichting Greenpeace Nederland and PAN Europe v Commission* after the previous judgement by the CJEU. [*Tweedale*, paras 80–88, *Hautala* paras 84-90]. One commentator[109] observed how the notion of 'environmental information' has been interpreted in light of the balance struck between openness and commercial interest when controversial types of information are at stake. While we agree with this note, some additional remarks are necessary.

The reasoning of the Court can be also read as a principle-based approach: first, the rationale underlying transparency and openness of risk assessors and managers is defined; then, a criterion of inclusion of information-types is identified; finally, individual information-types are included or excluded on the basis of this rationale. Such teleological interpretation has been provided by means of 'guidelines' [*Tweedale* para 93, *Hautala*, para 100] issued by the Court itself and

[106] Morvillo (2019)

[107] Morvillo (2019)

[108] Case C-57/16 P *ClientEarth v Commission* [2018] ECLI:EU:C:2018:660. In this case, the conflicting interests were represented by the public interest in disclosure of environmental impact assessment reports held by the Commission and the interest of the Commission itself in keeping a document secret during an ongoing decision-making process, a possibility granted by the exceptions provided by Art. 4(3) of Regulation 1049/2001. The former prevailed to prevent opaque external pressures over decision-makers. See Berthier (2016), para 75

[109] Morvillo (2019)

addressed to EFSA and the Commission. If this interpretation of Court's reasoning is correct, the question posed by one commentator[110] regarding the generalisability of Court's findings in *Tweedale* and *Hautala* can be answered by the extent to which the rationales of transparency and openness are capable of setting inclusive criteria and, in turn, information-types: the broader the scope of the two principles, the more data-types will be included.

To this extent, when comparing *Tweedale* and *Hautala* with their predecessor (*Commission v Stichting Greenpeace Nederland and PAN Europe*), it can be noted that the Court adopted a broader definition of openness. In the former two cases, openness is understood as broader principle that aims to promote democracy, accountability and trust; in the latter, the CJEU seems to not have discussed openness in general, rather narrowing its focus on proactive transparency measures in environmental matters.[111] Despite the different scope, the two approaches converge to one single criterion ('current and foreseeable emissions') that allows the inclusion of different information-types.

As this book analyses the general balance of public and commercial interests also in the context of reactive transparency, it is then necessary to 'learn the lesson' from the mentioned case-law on proactive measures and, in particular, the principle-based approach, while attempting to adapt it to the technical context and the amended legal framework of reactive transparency measures. Raw data, algorithms/source codes and operational data would certainly be needed in order to scrutinise the impact of regulated products on the environment, including emissions.

If this interpretation is correct, it could also solve the gap left open by commentators[112] when posing the doubt regarding the problem of 'irrelevant' data, whose obligation to disclose is questioned. It is argued that mandatory disclosure 'only relates to relevant data that can be extracted from the source of information and separated from other information contained in that source'.[113] Insofar such information (raw data, algorithms/source codes and operational data) is essential to foresee adverse health effects, an overriding public interest should be deemed to exist.[114]

It is incredibly hard to tell whether and when a similar case—in which environmental information is inferred from confidential data using probabilistic models and the access request includes operational data—would be brought before European

[110] Morvillo (2019)

[111] *Commission v Stichting Greenpeace Nederland and PAN Europe*, para 79: [i]t is apparent, in essence, [...] that the purpose of access to environmental information provided by that regulation is, *inter alia*, to promote more effective public participation in the decision-making process, thereby increasing, on the part of the competent bodies, the accountability of decision-making and contributing to public awareness and support for the decisions taken.

[112] Korkea-Aho and Leino (2017), p.1082

[113] Korkea-Aho and Leino (2017), p.1082

[114] This interpretation has also a strong epistemic relevance as the 'essentiality' of this information is also needed to prevent 'black box' problems in the Machine Learning algorithms deployed in this context. This dimension will be covered in Sect. 5.2.4 on Explicability

courts. If the ongoing datafication continues, it is likely that Courts will be forced to confront with these issues. To reason about to this eventuality, it is necessary to focus on openness and transparency also in the context of ongoing datafication trends.

2.2.3.3 *Arysta LifeScience Netherlands BV v EFSA* on Confidentiality Claims Decisions

Finally, a case—*Arysta LifeScience Netherlands BV v EFSA*[115]—is discussed to cast light on the judicial review of the decisions regarding the confidential treatment of commercially sensitive information by EFSA when the public interests to access information must be balanced with legitimate business interests to confidentiality.

In 2013, the Commission initiated the peer review of the active substance *diflubenzuron*, an insecticide used on apples, pears and mushrooms, and other crops [paras 25, 30]. In 2015, EFSA delivered its opinion and raised *a priori* concerns regarding the potential exposure to the substance named 'PCA (4-chloroaniline)' and 'its potential toxicological relevance for consumers, workers and residents or bystanders', also due to lack of data [paras 35–38]. On its own initiative, EFSA decided to publish its conclusions on its website. Before the publication, EFSA asked the applicant to identify possible confidential information in the conclusions, but eventually granted confidentiality status only to names of the authors of studies and reports, whereas it rejected the request grounded on the potential harm to applicant's commercial interests [para 39–40]. This choice was justified by the paucity of possible harm to applicant's image when confronted to its obligation to publish information likely to affect public health [para 41].

By means of five pleas, *inter alia* the applicant contested EFSA's decision to not to grant confidential treatment over the information for which it was requested. In particular, the applicant claimed that the scientifically incorrect assessment in question undermined its commercial interests and its reputation [para 105]. With this regard, EFSA noted that its conclusions were to be made available to the public for the protection of public health [para 106].

The General Court dismissed the action. While doing so, it provided a series of interpretative guidelines for the handling of confidentiality claims. First, the Court confirmed that 'confidential treatment of information [...] is the exception, whereas public access to that information is the rule', even when applicant's interest is legitimate [para 99]. Art. 63 of Regulation No 1107/2009 contains an 'open list' of information that might be subject to confidentiality claims, as long as the entity supporting them provides 'verifiable evidence showing in a concrete manner that the disclosure of that information might undermine his commercial interests' [para

[115] Case T-725/15 *Arysta LifeScience Netherlands BV, formerly Chemtura Netherlands BV, v EFSA* [2018] ECLI:EU:T:2018:977

108]. However, such possible harm cannot derive from the potential wrongfulness of the scientific conclusions [para 110].

The Court also recalled that Art. 63 of Regulation 1107/2009 promotes a balance between the transparency of review process and the protection of business secrets. To define what information falls within the remit of the notion of 'business secrets', the Court recalls the three criteria of the secrecy of information, the harmfulness of data publication and the objective worthiness of the interests underlying the non-disclosure of data. Such interests have to be weighed against the public interest that EU institutions' activity—including EFSA's risk assessment—are carried out 'as openly as possible'.[116]

Applicant's claim was rejected due to its failure to provide 'verifiable evidence' of the alleged threat to its business interests [para 130]. The Court also affirmed *ad abundantiam* the goodness of EFSA's decision on disclosing information on 'foreseeable health effects of the active substance at issue', also noting a settled jurisprudence that justifies the precedence of human health over economic considerations [paras 131–134].

General Court's final remark is the *trait d'union* between the present case and the other discussed in the previous sections. As with the cases of 'reactive' transparency, the Court adopted a principle-based approach—hence, first the identification of the rationale underlying transparency and openness, then the definition of a criterion to include information-types, and finally the selection of information-types—when evaluating the decision of an EU institution in the case of 'proactive' transparency measures. Differences and similarities, however, have to be noted.

On the one hand, the rationale underlying transparency in *Arysta* diverges from the one of the previous cases: data disclosure is justified by the protection of human health rather than the promotion of awareness and transparency in decision-making processes. One could argue that such an explanation is mainly due to the sectoral legislation at stake. However, this would be in conflict with the *obiter dicta* regarding openness from other cases, in which the Court had to confront with sectoral legislations as in *Arysta*. A more correct interpretation seems the one which suggests that, in *Arysta*, there was no risk-related 'decision'[117] involved, since the whole controversy took place in the assessment stages, which are usually prior to the actual 'decisions' taken by the risk managers, as it was in the other cases. Therefore, a general remark could suggest that the goals of transparency identified in the previous cases could not apply to the assessment scenario at stake in *Arysta*.

On the other hand, while the principle of transparency was grounded on a different rationale (protection of public health *vìs-à-vìs* transparency of risk management decisions), the identified criterion for solving the controversy is substantially

[116] Jaeger (2019)

[117] In the case at stake, the only decisions concerned are the ones taken by EFSA when opting for the publication of allegedly confidential information. This decision, however, occurred within risk assessment and should not be deemed to have managerial implications, i.e., the authorisation/denial to place the production on the market

the same, i.e., the foreseeability of health-related effects of regulated products. While in the previous cases such criterion was meant to be the gateway to include information among the ones to be published (*in concreto*), in *Arysta* it has been used as a justification of the balance struck between the public interest in proactive disclosure and commercial interests of undertakings by the European legislator (thus, *in abstracto*).

These findings leave open the question regarding how to interpret the principle of transparency in our domain. If, despite different underlying rationales, transparency is understood as a way to allow the foreseeability of health-related concerns by citizens both in reactive and proactive disclosure measures, this implies that, in light of the current datafication process, other informational components can be subsumed within the information needed for such foreseeability, not only in the context of reactive transparency measures—as argued above—but also for proactive transparency provisions. Forseeability seems also to be appropriate for dealing with risk assessment and risk communication: even though we noted that the rationale underlying transparency is different among the case law, its alignment in a 'transparency-as-foreseeability' interpretation seems consistent across all areas of risk analysis.

In conclusion, our analysis suggests that 'transparency-as-foreseeability' is a viable way to read the case-law of the CJEU in cases regarding the disclosure of environmental information. Such interpretation is mainly due to a principle-based approach that, following the identification of a rationale for transparency and openness, identifies a criterion to include certain information types (*rebus sic stantibus*, foreseeability) and proceeds with the inclusion of specific data-types. Following *Tweedale* and *Hautala*, the 'more transparency' approach proposed by the Legislator in the drafting of the 2019 Transparency Regulation can also be found in the recent jurisprudence of the CJEU, not only in the field of reactive transparency measures, but also when EFSA acts to divulge data or assesses confidentiality claims.

Will such convergence align with the new Transparency Regulation? I would cautiously suggest a positive answer. The overall rationale of the Regulation is, in essence, aligned with abstracts goals of ensuring trust in decision-makers by allowing scrutiny of their activities. However, different interpretations of the goals of transparency and openness, as well as an ambiguous use of these terms, might hinder the goals of promoting a unified approach towards 'more transparency'. Moreover, more speculative questions are left unanswered, especially regarding to (a) the changing nature of data ownership in this domain following the novel approach of the CJEU, and (b) the relationship between environmental information and algorithms in light of this case law.

2.3 A New Paradigm of Data Ownership in Food-Related Information

When analysing future trends in 2012, EFSA external reviewers[118] found that clarifying data ownership is a critical aspect, in particular when data is provided by Member States (p. 62). Reviewers noted lack of clarity on the ownership of data and once they are stored and made accessible in EFSA databases (p. 68). They concluded that contractual agreements with stakeholders might limit EFSA's ability to share data. The situation was not different in 2018, when reviewers noted 'in the face of data ownership by Member States, confidentiality claims of applicants and overall conflicting provisions in existing legislation both at national and European Union level'.[119] As we are now aware of *what* is data ownership (Sect. 2.1) and *why* it is critical, this section presents a wide range of topics and perspectives, all sharing a tentative answer to the question '*who* owns the data'?.[120]

Several authors have engaged the debate surrounding data ownership in this domain. Before providing new answers, it might be useful to understand how this debate has unfolded in the last years.

On the one hand, some authors have consistently argued that data ownership remains with the 'data originator'. *Ex novo* generations of data by EFSA are limited, thus data originators usually being third parties. This entails that, according to Kocharov,[121] data originators maintain ownership, whereas use, storage and release are regulated by EU law and contractual arrangements. In her view, the only way for EFSA to acquire ownership rights is under procurement contracts (e.g., confirmatory studies) for which EFSA retains data ownership. In the case of application dossiers, which are a major source of data for EFSA, companies maintain ownership due to the commercially sensitive nature of data—protected by confidentiality and data protection measures—and intellectual property rights.

We already noted that two kinds of measures—confidentiality and data protection—are in place to protect commercial interests of applicants or third parties. Raw data constitute a valuable asset for the companies engaged in R&D, hence the necessity of protecting this information against 'free riders' or unlawful accesses. The two sets of provisions are inspired by different rationales: confidentiality aims to prevent the disclosure of secret information, whereas data

[118] Ernst and Young External Evaluation of EFSA, Final Report 2012, available at https://www.efsa.europa.eu/sites/default/files/efsa_rep/blobserver_assets/efsafinalreport.pdf, accessed on 27/04/2022

[119] Ramboll and Coffey Third external evaluation of EFSA 2011-2016 Final Report http://www.efsa.europa.eu/sites/default/files/3rd-Evaluation-of-EFSA-Appendices100818.pdf, p. 12

[120] Two clarifications are necessary. First, the answers provided by these attempts are related to the technical and legal domain of food safety risk assessment; second, while this question is not explicitly included among our research questions, its preliminary assessment is necessary to formulate an answer to our second research sub-question

[121] Kocharov (2009)

protection guarantees a period in which other applicants cannot rely on previously submitted data unless agreed with the original data owner.[122]

When talking about the second sets of measures, Holle[123] links their existence to the exclusive right of reference to data for a certain period (e.g. 5 years). A critical point of this interpretation consists of the other mandatory requirement for data protection, i.e., the designation of data 'as proprietary' by the applicant. To maintain Holle's interpretation correct in light of this additional requirement, it can be argued that the designation of certain information as proprietary is a subjective condition that applicants claim *per se*, whereas the exclusive right to reference is an objective status recognised by the law. However, Simpson[124] argues that the meaning of 'exclusive right of reference' is often unclear and several alternatives are discussed.[125] While a shared interpretation of this criterion is still missing, Simpson's reading is consistent with the others when attributing ownership rights to applicants due to some kind of legal *status* attributed by the law.

Crucially, 'the law' is a purposeful simplification of a very complex legal framework that entails several pieces of legislations. In summary, applicants' ownership on raw data might originate from:

(a) The GFLR, as regards confidentiality claims over information-types set by Art. 39 discussed above;
(b) Food sectoral legislation, as regards information-types other than the ones disciplined by the GFLR;
(c) The Database Directive[126] which grants a *sui generis* protection for the maker of the database to prevent extraction and/or re-utilisation of data if a quantitatively/qualitatively investment has been made in obtaining, verifying or present the contents of the database (Art. 7).[127] As argued by Aplin,[128] ownership of the *sui generis* right lies with the maker of the database;[129]

[122] Holle (2014) and Simpson (2016)

[123] Holle (2014)

[124] Simpson (2016)

[125] *Inter alia*, an alternative wording is the "exclusive right of reference" criterion altogether with a requirement that applicants would have been able to "demonstrate ownership of the proprietary scientific or scientific data, by means of verifiable proof." Simpson (2016), see footnote 19

[126] Directive 96/9/EC of the European Parliament and of the Council of 11 March 1996 on the legal protection of databases [1996] OJ L 77/20

[127] Additionally, the Database Directive offers copyright protection for the selection and arrangement of the contents if they meet the requirement of being author's own intellectual creation (Art. 3)

[128] Aplin and Davis (2013), para 4.4.3

[129] My view of Aplin's assertion is that ownership of the *sui generis* right corresponds to ownership of the contents of the database, which is the object of legal protection in the *sui generis* regime

(d) Trade Secrets, as defined by Art. 2 of the Directive on Trade Secrets,[130] i.e., information that is secret in the sense that it is not, as a body or in the precise configuration and assembly of its components, generally known among or readily accessible to persons within the circles that normally deal with the kind of information in question; has commercial value because it is secret; has been subject to reasonable steps under the circumstances, by the person lawfully in control of the information, to keep it secret. As noted by an author,[131] proprietary data are treated in same way as trade secrets. EFSA practical arrangements for the implementation of the Transparency Regulation seen in Sect. 2.2 prescribe the secrecy of the information for which confidentiality can be requested as a necessary precondition for the claim.
(e) Contractual agreements. Data owners could restrict access to data do not follow under the scope of other means of protections granted by the Database Directive or trade secrets by means of contractual obligations. This additional possibility has been granted by the intervention of the CJEU in *Ryanair Ltd v PR Aviation BV*.[132] Kocharov[133] affirmed that contractual arrangements between the supplier of data and EFSA are secondary forms of regulation, for which EFSA is exposed to contractual liability.[134] The same can occur for physical or legal persons to whom data are disclosed, insofar they breach the licence terms. For instance, EFSA's first disclosure in *Hautala*,[135] a copyright notice was present in the cover letter introducing the release of data.[136]

The outcome of this system is a rather complex legal framework—not even considering the implementations of the aforementioned Directives in Member States which apply to national food safety authorities—that relies on several legal instruments for granting legal protection over valuable data. Some of these measures, like the exclusivity period for proprietary data and the *sui generis* protection for the contents of the database, present overlapping rationales, namely the goal of protecting the investment made in the gathering the protected data. Instead, the duration of such measures might range from 5 years for data protection measures,

[130] Directive 2016/943 of the European Parliament and of the Council of 8 June 2016 on the protection of undisclosed know-how and business information (trade secrets) against their unlawful acquisition, use and disclosure [2016] OJ L 157/1

[131] Simpson (2016)

[132] Case C-30/14 *Ryanair Ltd v PR Aviation BV* [2015] ECLI:EU:C:2015:10. For further comments see Aplin and Davis (2013), para 4.4.7 and Borghi and Karapapa (2015)

[133] Kocharov (2009)

[134] There is no limitation, however, to the raw data that need residual forms of protection as in the *Ryanair* case. Hence, all data qualify as potential object of protection by contract

[135] The correspondence between EFSA, Hautala and the other MEPs is available at https://www.asktheeu.org/en/request/is_glyphosate_safe_we_have_the_r, accessed on 27/04/2022

[136] 'All persons reproducing, redistributing, exploiting or making commercial use of this information are expected to adhere to the terms and conditions asserted by the copyright holder". The referred letter from EFSA can be found at https://bit.ly/3fHIf5X, accessed on 27/04/2022

to 15 years (renewable) for the *sui generis* protection, to the undefined time span of contractual arrangements. This might cause uncertainty towards the time span needed to proceed with data releases.

On the other hand, a minoritarian group of authors have identified weaknesses in this reasoning and claimed a different approach to the question of data ownership, by identifying the emergence of a new paradigm.[137] When discussing the concept of ownership, authors seem to argue that a simple distinction would be that companies are granted ownership by immaterial copyright rules, whereas authorities hold physical control, i.e., possession. While the EU agencies hold submitted information, the 'copyright of data'[138] belongs to companies, which can decide the conditions for data disclosure as seen in the case of replies to MEPs requesting access. The outcome is, in their view,[139] a 'strained situation of shared ownership'. This approach can be supported by other scholars who have referred to the 'collective intelligence' made possible by the fruitful interaction with users of systems used in food safety to foster accuracy and quality of the outcome.[140] Shared ownership might be understood as a necessary precondition of such collective intelligence.

Three major drawbacks—beside the *reductio ad unum* when discussing copyright issues only—affect this theory. On the one hand, such shared ownership shifts from a *monistic* (business operators) to a *dualistic* (EFSA *and* business operators) model. However, citizens then only play a marginal role in the view despite being the ultimate target of transparency mechanisms regardless of their rationale. On the other hand, while this approach is only focused on reactive transparency measures, a unified approach to proactive and reactive measures would be preferred also in light of the Transparency Regulation. Finally, the concept of possession is inherently inadequate to informational components, as we discussed when talking about ownership-related properties of data in Sect. 2.2. Despite these critical remarks, authors correctly challenge the idea of a monolithic approach to data ownership granted to companies by keeping into account the active role of EU agencies in determining *what* and *when* data can be disclosed.

While accepting the premise that a 'new paradigm of ownership' is emerging and in continuity with this trend, the approach proposed by our study takes it further by looking at the Transparency Regulation and the recent case law of the CJEU more holistically.

[137] Korkea-Aho and Leino (2017)

[138] As we showed above, narrowing the scope of legal protection to copyright would be, in principle, reductive

[139] Korkea-Aho and Leino (2017), p. 1086

[140] Alemanno and Gabbi (2016), p.213

2.3.1 Open Questions in Food Safety Data Ownership

Considering the possibility of a 'new paradigm of ownership', this subsection aims to identify new weaknesses in the traditional model in the light of the interplay between the existing norms, in continuity with the critiques highlighted by scholars remarking the inadequacy of the current conceptualisation of data ownership based on the concept of 'data originator'.

First, it has been observed that the publication of the non-confidential version of the application dossier is now the default transparency mechanism. According to our classification, the publication implies the shift of these units of information from private data (high rivalrousness, easy excludability) to some other form of ownership characterised by lower rivalrouness and harder excludability. It might be relevant to restate here that the publication of data by EFSA does not amount to an explicit or implicit permission or licence for the relevant data and information and their contents to be used, reproduced, or otherwise exploited in breach of any intellectual property right or data exclusivity rules, and the Union shall not be responsible for its use by third parties (new Art. 38.1a of the GFLR). With this regard, the ambiguity of Art. 38.1a has also to be remarked. Scientific data cannot be 'used, reproduced, or otherwise exploited in breach of any intellectual property right'. This appears to be in contrast with the new approach taken by Art. 3 of the 2019 EU Copyright Directive[141] when paving the ground for exceptions for text and data mining[142] made by legitimate users.

Several questions arise from this joint reading. For instance, how should we determine the concept of 'lawful access' when the amended GFLR does not confer any licence to use or reproduce IPR-protected data? The *lex specialis* criterion does not seem particularly helpful to determine the legitimacy of independent scientists willing to use data mining techniques over scientific information. Will food business operators be able to contractually limit the exploitation of data for legitimate text and data mining? A positive reply would seriously hamper the potential of automated literature reviews of their studies, whose growing importance has been largely discussed in the Introduction. Finally, who will technically restrict the possibility of lawful access for the 'security and the integrity of the networks and databases' where data are hosted (Art. 3(3) of the Copyright Directive)? This last point is particularly critical: while only rightholders are entitled to apply such measures, EFSA manages the infrastructure used to store the data and has the bear the burden of deciding the level of security measures, keeping into account rightholders' position while not acting on their behalf.

141 Directive (EU) 2019/790 of the European Parliament and of the Council of 17 April 2019 on copyright and related rights in the Digital Single Market and amending Directives 96/9/EC and 2001/29/EC [2019] OJ L 130/92

142 Art. 2(2) of the Copyright Directive 'text and data mining' means any automated analytical technique aimed at analysing text and data in digital form in order to generate information which includes but is not limited to patterns, trends and correlations for users having lawful access

While the disclosure of data to the public has been welcomed as an 'open data' approach in the public sphere, the doubts arising from the joint reading of these provision suggest the shift towards some kind of novel ownership model that is currently not possible to identify. However, the increased remit of exceptions in food legislation and in intellectual property law seems to prevent the attribution of a 'full' data ownership on the companies submitting data, contrary to what stated by the aforementioned scholars.

Second, we noted that, in the case of the procedure for the handling of confidentiality claims, several rounds of discussions between the Authority and the applicant may occur depending on the lack of agreement between the two parties. However, EFSA is now in entitled for taking decisions regarding the proactive disclosure of data and its conclusions are subject to the jurisdiction of the CJEU. This entails—following our definition of ownership—that EFSA enjoys a limited amount of data ownership, i.e., a form of *quasi*-ownership. Such *quasi*-ownership has to be granted to EFSA due to its active role in data-related decisions as the 'gatekeeper' of transparency, at least over the non-confidential and public version of the dossier for which EFSA takes decisions on the disclosure, including the 'if' and the 'what' has to be published. Moreover, it may be argued that, by providing so, the Legislator has purposely granted EFSA some kind of ownership rights to legitimately dispose of the data at the expense of the original data owner. The legal status of the commercial entity is somehow penalised by the expansions of EFSA's capabilities of disclosing data (emerging from the legislation) and the broad interpretation of the exceptions to confidentiality (emerging from the case-law). Therefore, even in this case a 'full' ownership is not enjoyable by the commercial applicant, nor can it be granted to EFSA.

Third, the reactive disclosure of environmental information is grounded on the 'foreseeability of environmental effects' criterion, justified either by the protection of human health or the need to increase trust in decision-makers. In the context at stake, the ability to foresee granted rights to citizens necessarily implies—at minimum—the access to data originated thanks to applicant's investment and protected as such. Such access rights are detrimental to the enjoyment of a 'full' ownership by commercial applicants: if an unlimited, unjustified and indiscriminate access to confidential data could theoretically be given to everyone, it means that even citizens enjoy some degree of ownership in potency or 'dynamically' (δύναμις, in the Aristotelic sense).

Fourth and finally, the ownership of the personal data has been largely overlooked by previous scholars. Confidential data also include findings generated through analyses that rely on recorded food consumption behaviours. In building a comprehensive ownership model, such human presence should ultimately be kept into account. According to the supporters of ownership-based theories of informational privacy, individuals may claim ownership over information regarding them, thus having recognised some form of *ius alios escludendi*.[143] In our context, however, this approach is unsatisfactory. While data subjects (i.e., surveyed individuals) might own their raw data, commercial applicants could claim ownership rights on the same information for the reasons seen above and both claims would be, in principle, valid. In practice, the endorsement of ownership-based theories in this context is likely to raise tensions between these parties. Moreover, if the possibility of granting access to individual data is guaranteed by means of reactive transparency measures, it looks plausible that further tensions between the public interest to process data and individual right to privacy might originate, ending in legal uncertainty as with the case of non-personal information and confidentiality.[144]

Taken together, these four considerations highlight some difficulties in adapting the law currently in force to one of the traditional models of ownership. Vice versa, ownership models do not fit properly to the data at stake if we consider them atomistically: while they might be appropriate for selected units of information in a given time, it is nearly impossible to use them to classify holistically the ownership of data at stake. Before trying to provide solutions to this issue, it might be interesting to challenge the interpretation of the EU law as regards the concept of environmental information as it has been provided by the ECJ.

2.3.2 Towards a New Concept of Environmental Information

In the light of our data-centric analysis, the outcome of the controversy between the 'foreseeable emission' vs 'purely hypothetical emission' criteria is particularly significant. Following the *Greenpeace* case, the ECJ has reached the conclusion that 'information on current and foreseeable emissions' constitutes environmental information for the purposes of the Aarhus Regulation and, in turn, Regulation 1049/2001. As such, it is subject to a restrictive interpretation of exceptions to disclosure justified on the basis of the protection of commercial interest.

In accordance with Art. 2(1)(d) of the Aarhus Regulation, the current definition of environmental information includes, in written, visual, aural, electronic or any other material form:

---

[143] A possible limitation of this approach is that all the parties involved in data protection activities also cover subjective positions typical of data protection law (e.g., data controller, data processor, data subject, third party, and so on). This reasoning, however, goes beyond subjective legal positions and should be read as a broader reflection on data ownership rather than in terms of a case-by-case analysis of data processing under a legal perspective

[144] Art. 4(1)(b) of Regulation 1049/2001 prevents public access to data and documents where disclosure would undermine the protection of privacy and the integrity of the individual. In the GFLR, we have observed a 'do not disclose' clause for the names of the individuals involved in toxicological studies, whereas nothing is said as regards food consumption data. In particular, specific criteria to (a) identify threats to informational privacy and integrity, and (b) balancing the public interest in accessing data and the individual right to data protection are missing

(i) the state of the elements of the environment, such as air and atmosphere, water, soil, land, landscape and natural sites including wetlands, coastal and marine areas, biological diversity and its components, including genetically modified organisms, and the interaction among these elements;
(ii) factors, such as substances, energy, noise, radiation or waste, including radioactive waste, emissions, discharges and other releases into the environment, affecting or likely to affect the elements of the environment referred to in point (i);
(iii) measures (including administrative measures), such as policies, legislation, plans, programmes, environmental agreements, and activities affecting or likely to affect the elements and factors referred to in points (i) and (ii) as well as measures or activities designed to protect those elements;
(iv) reports on the implementation of environmental legislation;
(v) cost-benefit and other economic analyses and assumptions used within the framework of the measures and activities referred to in point (iii);
(vi) the state of human health and safety, including the contamination of the food chain, where relevant, conditions of human life, cultural sites and built structures in as much as they are or may be affected by the state of the elements of the environment referred to in point (i) or, through those elements, by any of the matters referred to in points (ii) and (iii);

Considering the higher relevance of data analysis discussed in Sect. 1.3, not only are raw data becoming increasingly relevant in risk assessment, but also advanced analysis techniques contribute to the amount of studies that are carried out to assess the exposure to products with potential impact on the environment and human health. The ECJ stated that the overall goals of environmental-related provisions of Regulation 1049/2001—as interpreted in light of the Aarhus Regulation—are the involvement of citizens in decision-making processes, the accountability of decision-makers, and the promotion of public awareness of environmental matters.

The rationale inspiring such wide range of transparency measures is that access to data is decisive in the understanding of foreseeable effects of certain substances. For these reasons, any reactive disclosure of digital information shall include 'data that will allow the public to know what is actually released into the environment' [ECJ judgement in *Greenpace*, para 79] in particular when they allow a greater foreseeability of emissions and their effects on human health.

However, given the increasing relevance of Machine Learning algorithms, data only constitute one element of the whole picture. Today and most likely in the future, algorithms contribute to forecasting both emissions into the environment and their effects. This may entail that algorithms and the operational data about their accuracy (e.g., precision, recall, confidence scores, etc.) might fall under the remit of environmental information as far as they contribute to the general foreseeability of the emissions.

The definition of environmental information, as enriched by the contribution of the ECJ and its 'foreseeable emission' criterion, might also include new components, namely algorithms and data about their operations. Beside other justifications that can be provided under an ethical perspective,[145] a teleological reading of the concept of environmental information seems to open the doors for the possibility of enlarging its definition to the means that are used to generate predictions on the emissions. Similarly, the means used to forecast consequences of releasees into the environment for human health could be included on the same grounds among the information subject to disclosure measures.

2.3.3 Trust-Oriented Approaches to Data Ownership

Openness and transparency principles represent the cornerstones of EU food safety law. Therefore, underestimating their presence in the definition of a 'new paradigm of ownership' would lead to fallacious results. In light of the legal framework analysed so far, the principle of transparency is critically multi-faceted when it comes to its goals. Our analyses underlined that this principle is invoked to foster trust in food safety authorities, promote democracy, make EU food safety institutions accountable, protect human health, allow the public oversight of decision-making processes, and so on. While some of these goals are inextricably linked, other seems too far to be included with the operational remit of one single principle.[146]

It would therefore seem necessary to better identify some nuances of transparency that might fall within the remit of the concurring principle of openness when interpreting the GFLR.[147] It has to be preliminarily remarked that the distinction between the principles of transparency and openness is unclear. Understanding this relationship is not a mere exercise of legal interpretation, but it is crucial to correctly identify the scopes and the limits of transparency/openness measures in the newly amended food law. On the basis of their interpretation, authorities will take decisions regarding the extent to which data will be disclosed, possibly challenged before the CJEU which, in turn, will confront with the same interpretative matter if it decides to maintain its principle-based approach.

EFSA programming document 2020–2022[148] seems confused as regards the use of this terminology. First, it gives prominence to the principle of openness (p. 11). In other passages of the same document the two terms are used as substitutes ('transparency/openness', p. 13). Then, the paper justifies the publication of the

[145] See Explicability in Sect. 5.2.4

[146] Nothing would prevent us to attribute several meanings to the principle of transparency, hence my assumption of wrongfulness for this such-fits-all interpretation of transparency being possibly controversial. However, if we accept that transparency can be justified by such diverse grounds, we might end up in having no clue about the 'true' meaning of transparency (hence, the generation of legal uncertainty) and potentially contradictory interpretations when measures justified by the same principle do not synergize or clash with conflicting interests

[147] I have already attempted to discuss this topic in Sapienza (2019), but my conclusions were limited due to the ongoing legislative process at that time

[148] EFSA (2020)

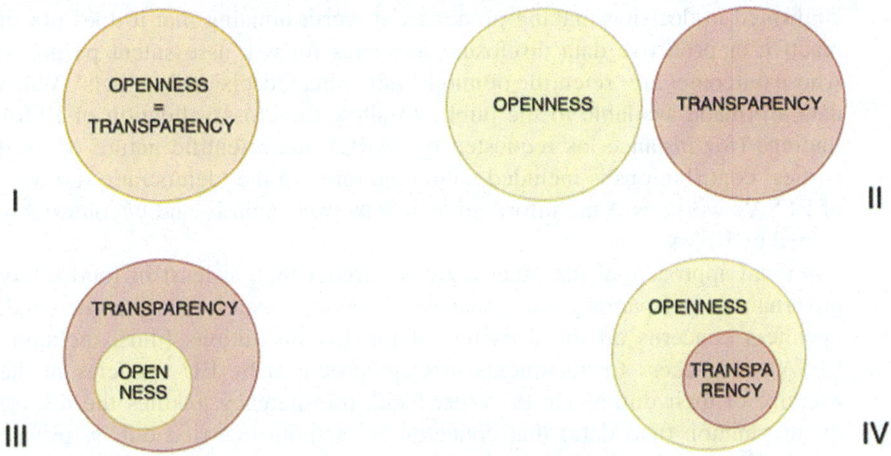

Fig. 2.2 Four approaches to openness and transparency of food safety data

non-confidential version of applications dossier as an 'openness' measure (p. 111). Finally, it refers to 'openness and transparency' as they were different (p. 130). To clarify their remit, let us now discuss four approaches to transparency and openness that can be found in the literature about food safety documents and data (Fig. 2.2).

I. First, openness and transparency are commonly used as perfect substitutes when describing 'good governance' measures indented to promote the deconstruction of layers of opaqueness and secrecy in EU decision-making.[149] It might be thus claimed that there is no major difference between the two concepts since they are both aimed at supporting democracy.[150] The CJEU has interpreted these two concepts as synonyms when discussing Art. 15 of the Treaty of Functioning of the European Union.[151] As already noted, in *Turco*, the CJEU stated that 'openness [...] contributes to strengthening democracy by allowing citizens to scrutinize all the information which has formed the basis of a legislative act' [paras 41–31, 59–61]. In *Bavarian Lager*,[152] the CJEU pointed out that Regulation 1049/2001 'is designed to ensure the greatest possible transparency of the decision-making process of the public authorities and the information on which they base their decisions' [para 49], thus implying little or no difference between the two terms. While this interpretation is largely

[149] Lodge (2003)

[150] Alemanno (2014)

[151] Consolidated versions of the Treaty on European Union and the Treaty on the Functioning of the European Union (TFEU) [2016] OJ C202/1. Art. 15(1) states that '[i]n order to promote good governance and ensure the participation of civil society, the Union institutions, bodies, offices and agencies shall conduct their work as openly as possible'

[152] Case T-194/04 Bavarian Lager v Commission [2007] ECR II-3201

confirmed in decision-making processes, it worth noticing that it does not fit reactive or proactive data disclosure measures for risk assessment purposes, whose outcomes are 'scientific opinions' rather than 'decisions' or 'acts'. When data are made available to the public to allow the cross-validation of EFSA findings (for instance, as requested by IARC), the scientific nature of third parties' contributions is included within the remit of the 'democratic scrutiny' of EFSA's work, as if the information at stake was entirely and uncontestably owned by EFSA.

II. A second approach, at the other extreme, argues that, at least in food safety governance, transparency and openness differ significantly. On the one hand, openness concerns all the activities of the EU institutions (thus, including EFSA) and access to documents that are produced by EU agencies in the exercise of their duties. On the other hand, transparency justifies the release of information (and data) that concerns the activities carried out by private parties.[153] In their view, the difference between the two principles is given by the data owner (public bodies vs private entities) and the nature of the documents to be released. However, we noted that, considering the increased weight given to access and distribution measures in the CJEU case law and in the Transparency Regulation, such monolithic approach to ownership is not always convenient, at least in light of the definition of data ownership that was given in the Introduction.

III. A third perspective may argue that transparency is the leading principle of food safety data disclosure measures and, therefore, it encompasses openness. This is due to multiple mentions of transparency (including the whole section II of the GFLR) in the provisions which mandate publications of data. Openness, instead, is used as synonymous for 'inclusiveness' and plays a marginal role only in scientific consultation of third parties[154] (Art. 32c and 32d of the amended GFLR) with no direct effects over information disclosure. Instead, it is worth reminding that EFSA's opinions do not consist of binding legislative acts,[155] but should be conceived as politically-neutral evidence-based research. The kind of inclusivity that one can imagine to be promoting democracy is not the same as the one of people's indirect participation to decision-making processes. Some scholars have argued that technical advancements and the culture of 'open data' have democratised science by giving citizens and stakeholders the possibility to intervene and make their voices heard.[156] While recognising that these individuals may speak 'louder' than in the past thanks to *openness-as-inclusivity* initiatives including open data, only qualified third parties can have sufficient expertise to actively participate the review of scientific evidence.[157] Therefore, 'open data' attempts might fail in their alleged

[153] Conte-Salinas and Wallau (2016), p.582

[154] Alemanno and Gabbi (2016), pp. 179–189

[155] Chapter 4 will discuss some challenges to this statement

[156] Alemanno and Gabbi (2016), p. 216

final goal, i.e., closing the gap between (perceived) technocratic EU institutions and citizens.

IV. A fourth possibility can be then considered. Openness might be a general principle discouraging secrecy and enabling EU institutions to have greater legitimacy and to be more effective and accountable to EU citizens. As we observed in *Tweedale* and *Hautala*, the CJEU has stated that the disclosure of scientific data contributes to an *open* (emphasis added) discussion, especially towards areas of scientific uncertainty and divergence, and fosters trust in EU institution [*Tweedale* para 75, *Hautala* para 60]. In this view, openness encompasses transparency, which, in turn, inspires sector-specific reactive or proactive measures—such as access to documents and data disclosures—that facilitate citizens' access to information held by EU institutions[158] in light of the general goals enshrined by openness. A major drawback of this interpretation is *mutatis mutandis* the same that affects transparency when encompassing goals that are undistinguishable.

To summarise, case law and academic literature have shown some degree of inconsistency in identifying the rationales underlying transparency and openness measures, the former being justified by the need of a democratic risk assessment, the latter by the necessity of ensuring collaborative forms of risk assessment. Neither of these approaches is satisfying. On the one hand, since risk assessment does not amount to a decision-making process, there is no necessity of supporting democracy in a scientific discussion. On the other hand, the number of individuals capable of actively contributing to the risk assessment is low.

Certain provisions mainly aim to foster a democratic oversight of the activities of EFSA by EU citizens, and in particular its relationship with commercial applicants; instead, different measures allow scrutiny over information for the cross-validation of scientific risk assessment results. Difficulties in these explanations also arise when these goals overlap (e.g., in the factual background of *Hautala* case). Can we then provide for an unified rationale for these goals while acknowledging their peculiarities?

Considering the low level of trust proved by the 2018 Fitness Check, the policy goals of the Transparency Regulation and the findings of the jurisprudence and the academic literature on transparency and openness, we could argue in favour

[157] Moreover, other scholars have argued against direct effects of democratic mutations in the 'knowledge economy' on the quality of democracy in a given society (Lynch (2016)), p.146. Democracy may be enhanced if citizens become active part of data-driven decisions, strategies and benefits that knowledge economy originates. See Durante (2019), p.187. Although our scenario is not directly linked to the knowledge economy given by Lynch, his approach to data and democracy seems appropriate to the context under discussion: even though 'open data' initiatives in agri-food safety risk assessment are qualified as democratic instruments, the mere inclusion—rather than active participation—of citizens in the data pipeline does not necessarily imply an increase in democracy. Different would have been the case of other 'open data' initiatives which might allow for a direct oversight of institutions (e.g., parliamentary open data)

[158] Alemanno (2014)

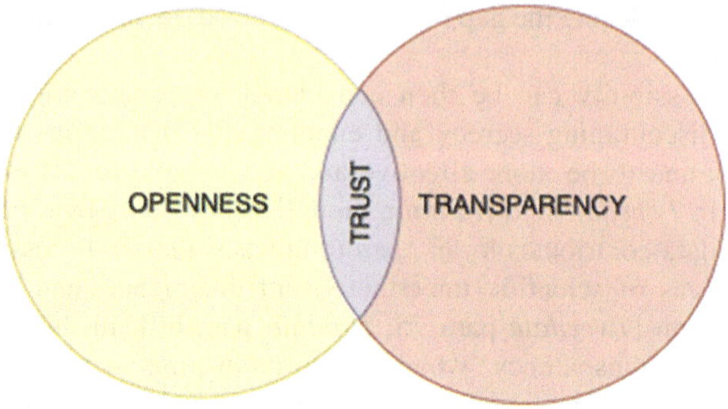

Fig. 2.3 A trust-oriented approach to openness and transparency of food safety data

of a model in which openness and transparency measures are diverse but partly overlapping (Fig. 2.3). In this model, actions that foster a democratic oversight over risk assessment procedures are closer to a 'transparency' rationale, whereas those promoting collaborative forms of risk assessment fall within the grounds of 'openness'. Both principles operate towards a common goal, which is represented by the promotion of trustworthiness of risk assessors. As regards openness, what is at stake is the reputation of EFSA as a scientifically sound institution that provides reliable opinions; with regards to transparency measures, they should foster trust by making public data held by EFSA to ensure independence and accountability of the Authority.

The general weaknesses of other models—beside the generation of confusing lexical choices—reflect difficulties in understanding the ultimate reason why heterogeneous stakeholders should have access rights to information that are held by other parties. This solution proposes trust as a unifying rationale.

2.3.4 Dynamic Ownership and Shared Benefits

The previous subsections have highlighted difficulties in answering to the question 'who owns food safety data' by relying on the existing legislation as interpreted by the ECJ. First, an in-depth scrutiny of the law has highlighted how the overlapping of different legislations prevent a straightforward answer to the question. Then, a focus on the concept of environmental information has shown how novel technical trends might broaden the concept of the concerned information to include algorithms used in the evaluation of foreseeable effects.

In sum, the conclusions drawn in the previous subsections suggest that (a) none of the entities that exchange information in the system—companies, authorities, citizens and surveyed individuals—enjoys 'full' ownership; (b) the discussion on data ownership shall be broaden to include other components, which are related to data, namely algorithms used to 'make sense' of data, in particular as regards the concept of environmental information; (c) trust can be used as the unified rationale to ground a new approach to data ownership in the light of the well-established principles of openness and transparency.

A possible starting point to answer the question 'who owns the data?' can be one of understanding the technical and legal context of food safety risk assessment data as a *distributed* form of ownership in which all the entities are granted rights of accessing, sharing, using or disseminating their data. These rights, however, are not unlimited and a constant balancing exercise—as in the case of EFSA's decisions on confidentiality claims—is needed.

EFSA plays the crucial role of mediating tensions between the other parties. The Authority does not attribute larger or smaller portions of ownership to each entity. Instead, it allocates distribution and access rights according to certain criteria identified by the GFLR as amended by the Transparency Regulation while following a principle-based approach as introduced by the CJEU. As the previous analysis suggests, transparency and openness are two core principles of the EU food safety legislation. If EFSA, as argued, is entitled to allocate ownership rights in a distributed framework, these principles shall inspire their decisions also under the new GFLR, also for reasons of continuity with the principle-based approach proposed by the CJEU.

Then, a clarification on the relationship between such unified model of distributed ownership and the trust-based theory of openness and transparency is needed to identify a common rationale that holds together different sets of measures allocating access and distribution rights. As remarked by the 2018 Fitness Check (*inter alia*, p. 10, 41, 45), trust is an essential component of the EU food system and, with this regard, it cannot be deemed to be external to the relationship among EFSA, the industry, and citizens, in particular when they share information. Under our data-centric and ownership-based discourse, the development of a trustworthy relationship between these actors is a foundational justification for distributing ownership rights when the measure granting access or distribution sets broad goals, as in the case of the GFLR or access to environmental information. Similar conclusions were reached in the context of an empirical, case-study analysis on different scenarios of data sharing and distribution practices—DECODE Project[159]—in which the absence of trust was deemed to increase the likelihood of a withdrawal from an information-sharing ecosystem.

Distributed ownership can also be deemed to increase trust. With more possibilities to data access, all the actors involved might enjoy the gain in trustworthiness that can occur following a novel conceptualisation of ownership. In fact, while citizens enjoy fewer obstacles to access data—reactively or proactively—EFSA and the food industry acquire trustworthiness before the public opinion.

On the one hand, EFSA can act in continuity with the spirit of the Transparency Regulation by strengthening its relationship with the citizens. On the other hand, as regards the industry, the possible loss in competitiveness of industrial players

[159] DECODE project (2020)

due to the wider availability of data for competitors would then be compensated by the climate of trust created by the possibility of accessing and reviewing data for cross-analysis.

2.4 Chapter Synopsis

This Chapter has extensively reviewed data ownership rules that are applicable to food safety. In particular, it has investigated the careful balance that has to be struck between the protection of the information generated by food business operators and its disclosure for an overriding public interest.

First, a discussion on data ownership has cast light on this topic. Different ownership models have been revised and the relevance of data transparency in food safety has been highlighted. Therefore, the preference for an ownership model that discards a straightforward ownership of food safety information on private entities only. Some authors have introduced the concept of a 'new paradigm of ownership'. While accepting its premises, the Chapter has sought for confirmation in the current legislative framework.

The Transparency Regulation has then been revised: new provisions significantly amend the existing legal framework by prioritising the disclosure. While transparency becomes the rule, confidentiality is limited to only certain elements whose harmful disclosure has to be proven by the food business operator. Findings from the ECJ case law confirm the preeminence of information disclosure over its confidentiality. The jurisprudence on environmental information and confidentiality requests management confirms that transparency prevails over secrecy. The Court also unfolded the several goals of transparency and openness, which range from democracy in public-decision making, public oversight, environmental reasons, inclusivity in scientific progress, and so on. The Court also left open the questions on algorithms: shall be they made available as they constitute the way in which we 'make sense of data' and, as such, should they be subject to the public scrutiny? The negative reply seemingly emerging from the rulings will be challenged.

Then, the Chapter ends with three proposals. First, algorithmic transparency—to be intended as the disclosure of source codes, log files, performance scores—is proposed to complement data disclosure measures. While an ethical justification, namely Explicability, will be proposed in Sect. 5.2.4, the chapter proposes to broaden the scope of information to be disclosed by means of a teleological interpretation of the norms in the light of the state of the art. Then, the Chapter introduces a distributed ownership model. While multiple legal instruments usually confer intellectual property over data to the food business operators, reactive and proactive measures suggest that the ownership of these data as defined in the Introduction shall be conceived as distributed rather than centralised. In fact, under the new legal framework, EFSA is in charge of attributing access rights by proactively or reactively distributing information as mandated by the law. The increasing remit of transparency and openness measures—above all, the publication by-default of the non-confidential version of applications dossiers and data—further decreases the monolithic interpretation of data ownership discussed by previous literature. To justify this novel regime, trust is proposed as the unified rationale underlying transparency and openness measures. This would eventually solve some conflicting interpretations and misused wordings in attributing meaning to the principles of openness and transparency.

References

Alemanno A (2011) Annotation of European court of justice, case C-79/09, Gowan comercio internacional E Servicos Lda V. Ministero Della Salute (Precautionary Principle)

Alemanno A (2014) Unpacking the principle of openness in EU law: transparency, participation and democracy. Eur Law Rev 1:72–90. ISSN 0307-5400

Alemanno A, Gabbi S (2016) Foundations of EU food law and policy: ten years of the European food safety authority. Routledge, Milton Park

Aplin T (2020) Research handbook on intellectual property and digital technologies. Research handbooks in intellectual property series. Edward Elgar Publishing Limited, Cheltenham. ISBN: 978-1-78-536834-9. https://books.google.it/books?id=jf3LDwAAQBAJ

Aplin T, Davis J (2013) Intellectual property law: text, cases, and materials. Oxford University Press, Oxford

Bartl A (2015) REFIT of food legislation: an opportunity to discuss implementation and enforcement issues. European Food Feed Law Rev 10:84–91

Bazylińska-Nagler J (2017) The right of access to environmental information in the light of the case C-673/13 P of 23 November 2016—European commission V Stichting greenpeace Nederland. Wroclaw Rev Law Administ Econ 7(2):66–82

Berthier A (2016) Transparency in EU law-making. In: ERA forum, vol 17, no 4. Springer, Berlin, pp 423–436

Beversdorf DQ, Roos RP, Allen Hauser W , Lennon VA, Mehler MF (2015) Animal extremists' threats to neurologic research continue: neuroreality II. Neurology 85(8):730–734

Borghi M, Karapapa S (2015) Contractual restrictions on lawful use of information: sole-source databases protected by the back door? EIPR 37(8):505–514

Borgman CL (2015) Big data, little data, no data: scholarship in the networked world. The MIT Press, Cambridge. ISBN: 978-0-26-202856-1

British Academy and Royal Society (2019) data-ownership-rights-and-controls-October-2018.pdf. https://royalsociety.org/-/media/policy/projects/data-governance/data-ownership-rights-and-controls-October-2018.pdf

Buonsante VA, Friel A (2017) What is information relating to emissions into the environment? Eur J Risk Regul 8(2):453–460

Conte-Salinas N, Wallau R (2016) The concepts of transparency and openness in European food law. In: International food law and policy. Springer, Berlin, pp 581–606

DECODE project (2020). Common knowledge: citizen-led data governance for better cities. https://decodeproject.eu/publications/common-knowledge-citizen-led-data-governance-better-cities

Duch-Brown N, Martens B, Mueller-Langer F (2017). The economics of ownership, access and trade in digital data; Digital Economy Working Paper 2017-01; JRC Technical Reports

Durante M (2019) Potere computazionale: L'impatto delle ICT su diritto, società, sapere. Mimesis

EFSA (2011) Activities, processes and quality assurance elements on data collection programmes with member states. EFSA Support Publ 8(3):127E

EFSA (2015a) Conclusion on the peer review of the pesticide risk assessment of the active substance glyphosate. EFSA J 13:4302

EFSA (2015b) Principles and process for dealing with data and evidence in scientific assessments. EFSA J 13(6):4121
EFSA (2015c) Response to open letter: review of the carcinogenicity of glyphosate by EFSA and BfR. http://www.efsa.europa.eu/sites/default/files/EFSA_response_Prof_Portier.pdf
EFSA (2015d) The EFSA data warehouse access rules. EFSA Supporting Publication
EFSA (2016) Open risk assessment: data. EFSA J 14:e00509
EFSA (2018) Administrative guidance for the processing of applications for regulated products (update 2019). EFSA Support Publ 15(1):1362E
EFSA (2020). EFSA programming document 2020–2022. http://www.efsa.europa.eu/sites/default/files/corporate_publications/files/amp2022.pdf. Accessed 26 April 2022
Faini F, Palmirani M (2018) The right to know and digital technology: proactive and reactive transparency in the italian legal system. In: International conference on electronic government and the information systems perspective. Springer, Berlin, pp. 164–174
Floridi L (2005) The ontological interpretation of informational privacy. Ethics Inf Technol 7 (4):185–200
Floridi L (2013) The ethics of information. Oxford University Press, Oxford
Fritschi L, McLaughlin J, Sergi CM, Calaf GM, Le Curieux F, Forastiere F, Kromhout H, Egeghy P, Jahnke GD, Jameson CW et al (2015) Carcinogenicity of tetrachlorvinphos, parathion, malathion, diazinon, and glyphosate. Red 114(2):70134–8
Gilsenan MB (2015) Data handling: observatories-databases-data storage-legal framework EFSA data collection. In: Options Méditerranéennes. series A: mediterranean seminars. CIHEAM-IAMZ, Zaragoza (Spain)-EFSA, European Food Safety Authority Parma
Global Open Data Index (2019) What is open data according to the open definition? https://index.okfn.org/faq/
Holle M (2014) The protection of proprietary data in novel foods–how to make it work. European Food Feed Law Rev 9:280–284
Jaeger M (2019) Access to file & confidentiality - the role of the court of justice of the European union. Accessed 26 April 2022
Kocharov A (2009) Data ownership and access rights in the European Food Safety Authority. European Food Feed Law Rev 4:335–346
Korkea-Aho E, Leino P (2017) Who owns the information held by EU agencies? Weed killers, commercially sensitive information and transparent and participatory governance. Common Market Law Rev 54(4):1059–1091
Lodge J (2003) Transparency and EU Governance: balancing openness with security. J Contemp Eur Stud 11(1):95–117
Lynch MP (2016) The internet of us: knowing more and understanding less in the age of big data. WW Norton & Company, New York
Mahmoud S (2019) Sharing and caring. IPPR Progr Rev 26(1):78–89
Malgieri G (2016) Property and (intellectual) ownership of consumers' information: a new taxonomy for personal data. Privacy in Germany-PinG (4):133–150
Malgieri G (2018) User-provided personal content'in the EU: digital currency between data protection and intellectual property. Int Rev Law Comput Technol 32(1):118–140
Martin EA (2009) A dictionary of law. OUP Oxford, Oxford
McHenry LB (2018) The Monsanto papers: poisoning the scientific well. Int J Risk Safety Med 29(3–4):193–205
Microsoft (2019) Configure the user connections server configuration option - SQL server | Microsoft Docs. https://docs.microsoft.com/en-gb/sql/database-engine/configure-windows/configure-the-user-connections-server-configuration-option?view=sql-server-2017andviewFallbackFrom=sql-server-2017%7B%5C%%7D7D
Morvillo M (2019) The general court orders disclosure of glyphosate-related scientific studies: Tweedale, Hautala, and the concept of environmental information in the context of plant protection products. Eur J Risk Regul 10(2):419–427
Moules R (2017) Significant EU environmental cases: 2016. J Environ Law 29(1):177–188

Ostrom E, Calvert R, Eggertsson T (1990) Governing the commons: the evolution of institutions for collective action. Political economy of institutions and decisions. Cambridge University Press, Cambridge. ISBN: 978-0-52-140599-7. https://books.google.it/books?id=4xg6oUobMz4C
PAN Europe (2020) Fraud in German laboratory casts additional doubts on the 2017 re-approval of glyphosate and on the entire EU pesticide safety evaluation procedure. https://www.pan-europe.info/press-releases/2020/02/fraud-german-laboratory-casts-additional-doubts-2017-re-approval-glyphosate. Accessed 26 April 2022
Peel J (2012) Of apples and oranges (and hormones in beef): science and the standard of review in WTO disputes under the SPS agreement. Int Compar Law Quart 61(2):427–458
Portier CJ (2015) Open letter: review of the carcinogenicity of glyphosate by EChA, EFSA and BfR
Portier CJ, Armstrong BK, Baguley BC, Baur X, Belyaev I, Bellé R, Belpoggi F, Biggeri A, Bosland MC, Bruzzi P et al (2016) Differences in the carcinogenic evaluation of glyphosate between the international agency for research on cancer (IARC) and the European food safety authority (EFSA). J Epidemiol Commun Health 70(8):741–745
Rodotà S (2015) Il diritto di avere diritti. Gius. Laterza/Figli Spa
Rusconi G (2016) Food safety and policy in the european union. In: International food law and policy. Springer, Berlin, pp 451–483
Sacco R (1991) Legal formants: a dynamic approach to comparative law (Installment I of II). Amer J Compar Law 39(1):1–34
Samuelson P (2000) Privacy as intellectual property? Stanford Law Rev 52:1125–1173
Sapienza S (2019) Transparency and openness in food safety: insights about new data confidentiality rules. https://doi.org/10.5281/zenodo.2766359
Simpson C (2016) Data protection under food law post: in the aftermath of the novel foods regulation. Eur Food Feed L Rev 11:309
Spedicato G (2016) Digital lending and public access to knowledge. In: Intellectual property and access to im/material goods. Edward Elgar Publishing, Cheltenham
Tamò-Larrieux A (2018) Mapping the privacy rationales. In: Designing for privacy and its legal framework. Springer, Berlin, pp 27–43
The Guardian (2017) EU report on weedkiller safety copied text from Monsanto study | Environment | The Guardian. https://www.theguardian.com/environment/2017/sep/15/eu-report-on-weedkiller-safety-copied-text-from-monsanto-study. Accessed 26 April 2022
von Holleben H (2013). Judgment of the general court of the EU on access to information under substance law: case T-545/11, Judgment of 08 October 2013. Eur J Risk Regul 4(4):565–578
Wolf S (2013) Access to EU environmental information: EU compliance with Aarhus Convention. In: ERA Forum, vol 14, no 4. Springer, Berlin, pp 475–491

Chapter 3
Food Consumption Data Protection

3.1 Food Consumption, Surveys and Mobile Applications

This section discusses the implications of food consumption data processing. This information is crucial for food safety risk assessment as it constitutes the necessary benchmark to evaluate long-term effects of possible food-related hazards. Several interrelated issues can be linked to its processing, including its status under legal data protection rules, the general level of compliance, and the collective implications of dietary intake analysis as regards groups. The section will be finalised by discussing why the protection of this information-type deserves attention in the academic debate and in practise for the implications on human health.

Section 3.1 introduces the discussion on food consumption data processing by referring to current practices in data collection and analysis, also beyond food safety. As it will be discussed, novel collection methods show a growing interplay between traditional methodologies and emerging trends.

Section 3.2 contextualises our discussion on dietary intake information to some key areas of the general debate about data protection. As it will be highlighted, food consumption data are placed in-between certain legal categories. Moreover, a group-level discussion is introduced.

Section 3.3 clarifies the relationship between food consumption data and the datafication of personal identity. On the basis of their connection, the relevance of *fairness* is introduced as regards the implications of food consumption data analysis with regards to groups that correspond to certain dietary preferences.

3.1.1 Food Safety Dietary Intake Surveys

EFSA Food Consumption Database is the portion of the Data Warehouse that contains information on the amount of foods eaten by European citizens. While

S. Sapienza, *Big Data, Algorithms and Food Safety*, Law, Governance and Technology Series 52, https://doi.org/10.1007/978-3-031-09367-8_3

it follows the Data Warehouse rules mentioned in the previous chapter, access to this information is granted to companies willing to place regulated products on the market and to the public.

EFSA has the right to use the raw, individual food consumption data for carrying out risk assessments and other scientific analyses. An authorisation from the data provider (usually, a Member State) has to be obtained prior to the further processing of data. Summary statistics (high level of aggregation) from the Comprehensive Database are made available to the public on EFSA's website and can be used by applicants submitting requests to place regulated products on the market to support their claims.[1] The majority of occurrence data sent to EFSA comes from laboratories involved in national monitoring programmes and are submitted to EFSA by national competent authorities in EU Member States. Other data providers include the food industry (mainly via networks), universities, consumer associations, and the European Commission (DG SANCO). EFSA has established several data collection networks, composed of representatives of national competent authority data providers, to support its data collection activities in the field of food consumption data.[2]

The Comprehensive Database contains data characterised by high granularity, both as regards food ingredients and individuals.[3]

First, food consumption input data are classified according to the age of the individual consumer, the following population classes.

- Infants: <12 months old
- Toddlers: ≥12 months to <36 months old
- Other children: ≥36 months to <10 years old
- Adolescents: ≥10 years to <18 years old
- Adults: ≥18 years to <65 years old
- Elderly: ≥65 years to <75 years old
- Very elderly: ≥75 years old

After several pilot studies (e.g., PANCAKE and PILOT-PANEU[4]), the chosen methodology for all age groups (with the exclusion of infants and toddlers) is a 2-day non-consecutive 24-hour food dietary recall, i.e. a survey intended to gather data about the food and beverages consumed in the previous 24 hours.[5] This interview might be self-administered or conducted by an expert. EFSA has reported that interviews last for 30 minutes on average. When the interview is carried out by a nutritionist, it consists of a computer-assisted personal interview (CAPI) or via telephone (CATI). Validated dietary softwares are used to carry out the survey, for

[1] EFSA (2011)

[2] Gilsenan (2015)

[3] EFSA (2019)

[4] See, respectively, Ocké et al. (2012) and Ambrus et al. (2013)

[5] EFSA (2014)

instance to display pictures of foods to be chosen by the surveyed. Alternatively, picture books are used as visual supporting materials during the talk.

Background information including age, sex, marital status, region, rural/urban area, size of the household, household income, employment status is collected to detect dietary patterns. Instead, specific, predefined dietary patterns, whether through personal choice (e.g., vegetarians) or because of health conditions (e.g., diabetes or coeliac disease) are recorded. Finally, two measures (body weight and height) are directly used to perform exposure assessment.[6]

Since 2019, foods are recorded as raw primary commodities (RPCs) (i.e. single units of unprocessed, harvested or slaughtered foods, like "apple"), RPC derivatives (i.e. processed RPCs, like "apple juice") and composite foods (i.e. foods consisting of multiple components, like "apple strudel").[7] Composite food consumption represents a unique challenge, as it has to be disassembled into RPCs. To do so, three tables are used. First, consumption data for composite foods (input data) are disassembled into RPCs and RPCs derivatives (intermediate data) using both the disaggregation table and the probability table. Then, processed RPCs are converted in RPC by a conversion table (output data).

The use of disaggregation table is meant to fragment composite food into derivative RPCs or RPCs. Hence, when a composite food has different flavours (e.g., type of muffin—chocolate or plain) or components (e.g., type of meat in a meatball), this table is also used to probabilistically assign such components. Composite foods components are manually entered into the table by selecting the corresponding recipes (commercial products are indexed via Mintel Global New Products Database, while recipes are gathered from the website allrecipes.com). Then, a probability is calculated on a frequency indicator based on components' description.

Probability table, instead, is helpful to obtain more granular data. For instance, consumption records for vegetable oil will be assigned to a more specific oil type (e.g., sunflower oil, olive oil) based on probability on consumption of age classes (i.e. subjects < 1 year old, subjects ≥ 1 to < 10 years old and subjects ≥ 10 years old). This is expressed as a number ranging from 0 to 1.

Finally, a conversion table translates quantities of RPC derivatives into their equivalent weight of RPCs before processing, i.e. the original ingredients. The conversion table is based on several papers and relevant literature, while no probabilistic method is involved.

[6] EFSA (2014), ch. 7
[7] EFSA (2019)

Bulding the Comprehensive Database: A Practical Guidance EFSA[a] reports the following example. For subjects aged ≥10 years, 2215 individual consumption records were classified as "Pastries and cakes" in the Comprehensive Database. According to the probability table these consumption records may either be assigned to "Buns" (i.e. small pieces of sweet bread) with a probability of 50% or to "Sponge cake" with a probability of 50%. This has resulted in 1100 of these consumption records assigned to Buns and 1115 of these consumption records assigned to Sponge cake. The Comprehensive Database contained a consumption record where 175 grams of Sponge cake were consumed by an individual. According to the disaggregation table such Sponge cake is made of 26.1% egg, 24.5% wheat flour, 24.5% sugar, 24.5% butter and 0.4% baking powder. This has resulted in the following consumption records of 46.6g egg, 42.9g wheat flour, 42.9g sugar, 42.9g butter and 0.7g baking powder for the individual.

[a] EFSA (2019)

Attempts have been made to lessen the burden of surveyed consumers while reducing the number of self-reported foods needed to infer compliance with dietary recommendations trough a decision tree classification algorithm. A study showed that age, sex and consumption data about 113 foods on the 3911 available in the UK National Diet and Nutrition Survey (3%) are needed to predict compliance with food guidelines[8] with 72–83 % accuracy depending on the food category. A similar study[9] managed to increase the accuracy of 2.5% on average by including nationality and marital status and reducing the number of foods. Despite not being related to risk assessment, the inclusion of socio-demographic characteristics regarding parents and guardians has been required to perform a study on child consumption of foods and obesity, alongside children's physical activity data.[10] In the fields of food consumption data analysis, dietary patterns are a consolidated trend.[11] Clustering and Principal Component Analysis (PCA) has been used to generalise food consumption for pattern identification.[12] Results have then been processed through Reduced Rank Regression (RRR) algorithms to correlate patterns and blood samples with diabetes.[13] Similarly, nutrition patterns have also been proven to be successful co-predictors cardiovascular risk prediction.[14]

[8] Giabbanelli and Adams (2016)

[9] Rosso and Giabbanelli (2018)

[10] Lazarou et al. (2012)

[11] Hu (2002)

[12] Hearty and Gibney (2013)

[13] Batis et al. (2016)

[14] Rigdon and Basu (2019)

3.1.2 Food Diaries, Data Mash-Up, Mixed Datasets

Mobile food diaries are applications (software) capable of collecting, analysing and sharing self-assessed diary intake information using mobile phones. Their functioning can be summarised as three kinds of components (namely essential, additional and advanced features) depending on their design. In this section, all the percentages refer to the number of apps having such features according to the analysis conducted by Maringer et al.[15] ($n = 176$) on the UK app marketplaces (Fig. 3.1).

Essential features can be described as the vital components of every food diary. They consist of a search box to query a food database (74%). Less advanced food diaries do not perform any research on the Web, thus relying on built-in databases. The output consists of best-matching foods (eventually, together with their brand), accompanied by macro-nutrients information. When the user manually adds each meal to the diary, nutritional data, reports, and analyses are displayed. In less advanced apps that only rely on essential features, dietary data is not shared with any third party, nor users build profiles by providing additional, non-food related information.

Additional features are popular components of food diaries, which provide a more user-friendly usage of the app. Mildly-advanced apps feature at least one of them. They include the possibility of building a profile by adding other personal data such as weight (43%), BMI (13%), level of activity (61%), number of steps taken (11%). Such profile is directly correlated to the food diary, thus increasing the ability of keeping track of user's health (e.g., users can achieve a better weight management by confronting burned and ingested calories). Furthermore, a popular additional feature is performing research over a database system. For instance, diet app MyNetDiary calculates calories from users' manual search in a large database (894,000 foods records) and allows users to keep track of their intakes. Generally, API-based research is performed. As a result, the research output is usually up-to-date and correlated with user-contributed food information gathered from nutritional labels on products.

Advanced features include wearable connections, image recognition and data sharing with third parties. These components are certainly not necessary to monitor diary intake, but deeply enhance user experience. Within the 'profile' section, it is possible to connect the food diary to a wearable device (e.g., smartwatches, fitness bands, etc.) to gather real-time information on calories burned, number of footsteps, heart rate, and so forth (14%). Certain food diaries feature image recognition to perform the food research. They may allow either a barcode-based query (22%) or food image recognition (25% for labeled or packaged food products, 12% for 'pure' food

[15] Maringer et al. (2018)

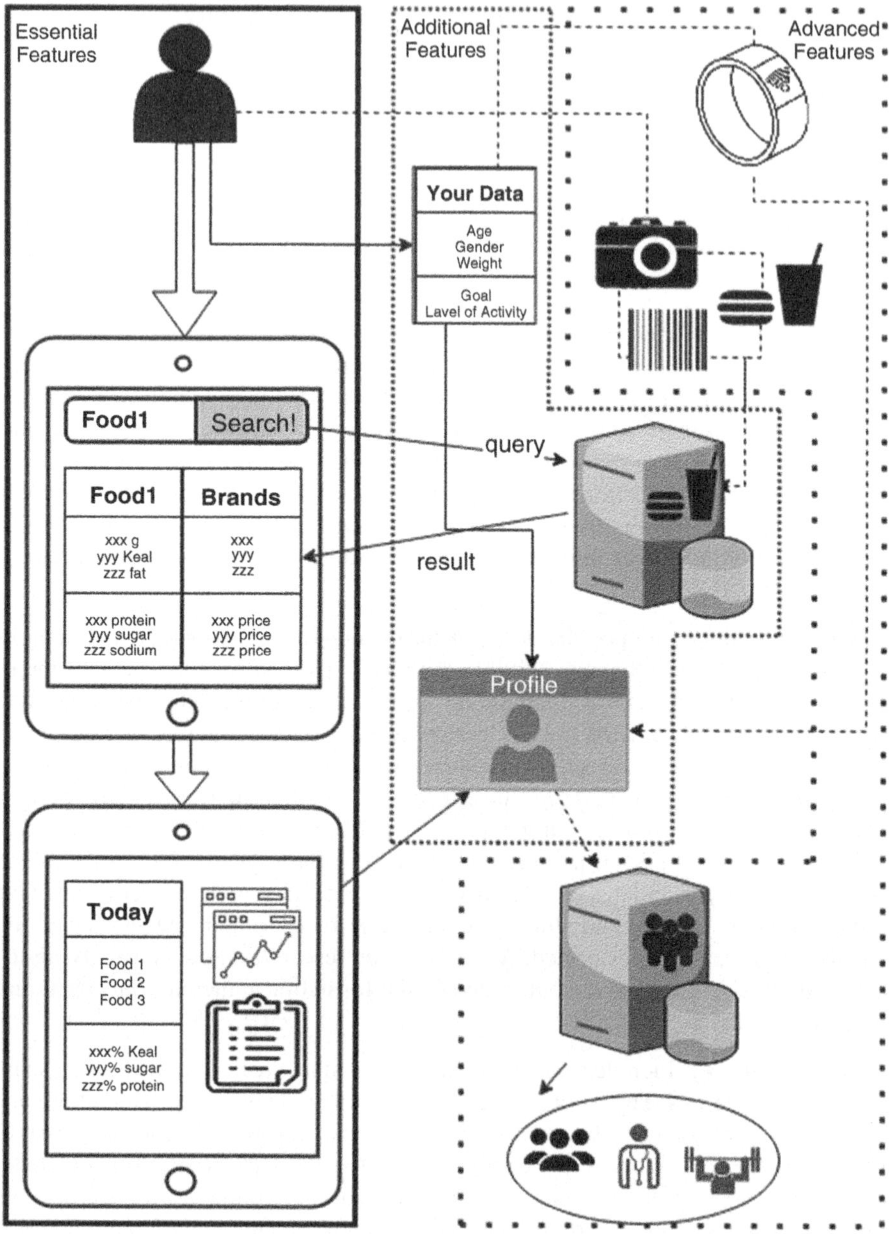

Fig. 3.1 Data flows within a mobile food diary. Essential features are connected by larger arrows; additional features are linked by standard-side lines; advanced features are bridged by dotted lines

images) using deep learning.[16] Such pictures might also be included in the profile section to maintain a visual record of consumed food. Lastly, 'social' advanced features allow data sharing with third parties. First, users can share their diet plans and other information in a networked environment (13%), which might include a certain amount of gamification (badges and awards for more active/healthier users). Secondly, users might be connected to doctors and nutritionists (30%), who can access their data to provide consulting. Thirdly, consulting might be offered by a personal trainer (11%), if the user prefers a more fitness-oriented experience. Lastly, profile data might also be included in data aggregators, such as Apple HealthKit and GoogleFit, and so forth (25%).

Findings from this analysis highlight the possibility of combining heterogeneous information-types and generate "data mash-ups". This trend is not exclusive to food consumption data and has been analysed by the EU Commission in the Guidance on the Regulation on a framework for the free flow of non-personal data in the European Union.[17] It has been observed that "[a] mixed dataset consists of both personal and non-personal data. Mixed datasets represent the majority of datasets used in the data economy and are common because of technological developments such as the Internet of Things (i.e. digitally connecting objects), artificial intelligence and technologies enabling big data analytics".

This short—yet, highly authoritative—statement regarding the collection of several data-types has implications in our scenario as regards, in particular, what information can be derived or inferred from food consumption data when other sources are used and what legal regime(s) govern the processing of such novel information.

Smartphones may have a significant impact on food consumption data gathering. Starting from EFSA Strategy 2020, the Authority has progressively endorsed the involvement of citizens through collaborative platforms and data crowdsourcing, both to foster trust in the Authority by engaging with citizens and to gather updated data. For instance, a 2017 EFSA tender (OC/EFSA/AMU/2017/02) asked participants to provide a "prototype design of a mobile app" to collect information pertaining to infants' consumption data, alongside parents' personal data regarding age, sex, region, size of the household, education, recent employment status and professional category.

Food consumption data gathered through memory-based methodologies (i.e. the traditional interviews and surveys described above) have been defined as "pseudoscientific" by a highly critical research.[18] Similarly (but with much less emphasis), Van den Puttelaar and colleagues[19] detected several disadvantages both

[16] Marin et al. (2019)

[17] Communication From the Commission to the European Parliament and the Council "Guidance on the Regulation on a framework for the free flow of non-personal data in the European Union" COM(2019) 250 final

[18] Archer et al. (2015)

[19] Van den Puttelaar et al. (2016)

for 24-hour food dietary surveys and questionnaires: long and time-consuming interviews, recall biases, likelihood in forgetting consumed food, and lack of contextualised information have been identified as the major drawbacks for dietary surveys.

One of the proposed solutions consists of using Internet and smartphones for dietary assessment, similarly to pivotal studies conducted in the USA in 2010.[20] This approach is deemed to have the potential both to lower costs and burden and to facilitate a timely and accurate data analysis.

A good number of examples can be found in the literature. Authors[21] have modelled a mobile application—"FoodProfiler"—to overcome the aforementioned issues of memory-based methodologies. The mobile app proposes a 2-hour, self-executed and game-based food consumption recall. Other scholars have reviewed the use of publicly available apps for self-documenting food consumption,[22] despite some remarks on the reliability, the quality and the interoperability of data for scientific uses.[23]

EFSA has seemed cautious in endorsing collaborative and smartphone-based methods, hence more research—outside the scope of this document—is needed to assess the validity of self-documented food consumption data collection. A general trend, instead, has to be noted. Despite EFSA's caution, single studies and extensive reviews show the promising results of these gathering methods to ease dietary intake assessment and data quality, especially in comparison to the methodologies currently in use by EU Member States competent authorities. In sum, despite being far from perfect, these methods could nonetheless provide significant benefits for dietary intake data gathering in the near future. This paradigm shift, however, is not exempt from significant concerns regarding the nature of the data at stake and will be discussed in Sect. 3.3.[24]

[20] Thompson et al. (2010)

[21] Van den Puttelaar et al. (2016)

[22] Eldridge et al. (2019)

[23] Maringer et al. (2018)

[24] An other crowdsourcing to be used for reference, smartphone-based solution supported and funded by EFSA is Mammal (see https://mammalnet.com/collaborators/, accessed on 27/04/2022) project, which consist of a smartphone app (IMammalia) through which EFSA "will use collected information on abundance and distribution of wild mammals in order to assess the risk of diseases affecting wildlife, livestock, and humans". The collection of such user-generated information can be considered as part of the ongoing process aiming at fostering collaboration between the Authority and the general public, generating trust in EFSA and accessing update information with lower burdens

3.1.3 *Food Consumption Data, Derived and Inferred Information*

A taxonomy provided by the Information Accountability Foundation[25] and publicly endorsed by the UK Information Commissioner's Office[26] outlines that personal data can also be described by looking at the their origin: *provided* (i.e. originated by a conscious individual), *observed* (i.e. recorded by a third party), *derived* (i.e. extracted from other data in a mechanical way) and *inferred* (i.e. extracted from other data following a probability-based analysis).

The life-cycle of dietary information in food safety risk assessment departs from provided or observed records of food consumption. Then, this information is aggregated into age clusters for further analysis. Finally, it is juxtaposed to non-personal data to infer possible exposure levels to pathogens and, eventually, negative effects for human health. Dietary records are subject to transformations that affect the legal regime to which they are subject. It can be preliminary observed that, despite the apparent fragmentation of dietary information processing under this taxonomy, certain "safe harbours" in data protection law provide for sufficient certainty as regards the legal regime.

Legal boundaries in the information life-cycle guide our understanding of the regime that governs the data processing. First, at the moment of collection (provision or recording), dietary intake information usually qualifies as personal data. From a legal perspective, this occurs every time it is linked to an identified or identifiable individual, which is the usual case in practice. In fact, metadata from EFSA[27] show that individual names are replaced by an identifier that is not disclosed to the public.[28]

Following the aggregation step, data might no longer be linked to an identified or identifiable individual, thus falling outside the scope of the GDPR when used in ways that do not trace back to persons. This aggregation-anonymisation equation is not always correct, as public authorities competent in the field of data protection—including in European ones (see below)—have stated.[29] However, derived information might also be extracted by observing dietary *patterns*, or sequences of recorded consumptions. For instance, when observing regularities in the consumption of certain foods, for instance, the complete absence of meat or fish, it is possible to associate the pattern to the vegetarianism of the observed consumption. The same goes with other personal attributes, such as the presence of food-related conditions

[25] Abrams (2014)

[26] ICO (2017)

[27] Available at https://doi.org/10.5281/zenodo.1215993, accessed on 27/04/2022

[28] In the description of these metadata, EFSA reports that "data protection" yes/no indicators show whether the structural metadata element will be published or not (yes = will not be published, no = will be published). ORSUBCODE is the Unique subject identifier and seems to be covered by data protection under Regulation 45/2001 (now replaced by Regulation 2018/1725)

[29] FTC (2015) and PDPC (2018)

(e.g., coeliac disease or gluten sensitivity), religious belief (e.g., fasting) or market preferences (e.g., online shopping). From a legal perspective, the correctness of the derived information does not hinder the qualification of this data as personal, insofar it is still linked to an identified or identifiable individual.

While the generation of derived information does not require other background data to be generated, inferences are usually made on the basis of multiple variables for performance reasons. This is necessary in food safety risk evaluation because exposure assessment measures the likely level of hazard on the basis of food eaten in a given area, at a given time, by a given population. Another significant difference between derived and inferred information concerns their nature: on the one hand, derived information is always deterministic and the same input generates the same output; on the other hand, inferences are stochastic or probabilistic by nature, which entails that results are expressed in forms of a probability of correctness.

In summary, food consumption data can be used as *proxy variable*. In statistics, proxy variables or surrogate variables represent a "measurable variable that is used in place of a variable that cannot be measured" or "a variable that can be measured (or is easy to measure) that is used in place of one that cannot be measured (or is difficult to measure)".[30]

3.2 Dietary Information Processing and Data Protection Between Law and Ethics

As illustrated in the previous section, food consumption data can reveal a significant number of other attributes concerning the data subject. Moreover, background information is, in general, not exempt from risks.[31] However, these risks emerging from this data processing are balanced by the adoption of pseudonymisation techniques by EFSA,[32] a suitable safeguard measure according to Articles 13, 27, and 33 of Regulation 2018/1725. It is worth noticing that Member States adopted jeopardised data protection measures when building national datasets. For instance, Spain simply treated personal data "as confidential".[33] France had to obtain a mandatory authorisation by the French Data Protection Authority, while

[30] Upton and Cook (2014)

[31] Recital 46 of Regulation 2018/1725 specifies the extent to which certain data processing activities might results in risks for individuals. *Inter alia*, risks could emerge from the processing of children's personal data, a large amount of personal data or a large number of data subjects. These three conditions are met in food consumption and background information processing

[32] EFSA et al. (2019), p.13, p.19

[33] Marcos et al. (2016), para 2.6. In the context of food safety risk assessment data, the use of such wording might be misleading. As we noted, confidentiality regime applies to information submitted in the context of regulated product and confidentiality is granted after the scrutiny illustrated above. It might be the case that "as confidential" is a communicative substitute for "in compliance with data protection law" that has been used when informing participants on matters related to data

the Netherlands did not report about data protection aspects,[34] Italy stated that "[t]he survey was exclusively observational and non-invasive, ethical aspects were related only to the collection of information on food habits that may be related to health and thus might be sensitive. INRAN is part of the National Statistical System (SISTAN) and guarantees individual data protection. An additional ethical committee review of the study protocol was considered unnecessary".[35] Remarkably, datasets generated by these studies and made available to EFSA include both dietary intake and background information, as established by the official guidelines.[36]

From EFSA metadata,[37] other data elements are of particular interest. EFSA seems to collect the "self-defined ethnic group" of the surveyed individual, her or his "special conditions" and whether the subject follows a "particular eating pattern". These information-types might fall within the remit of Article 10 of Regulation 2018/1725 and be processed as special categories of data if they express the ethnic origin, health conditions or dietary patterns reflecting religious or philosophical believes and if they are recorded as such. For instance, a surveyed individual might declare to adhere to Islamic dietary laws. This would entail, among others, prohibitions regarding alcohol or pork. Differently, her or his recording can show adherence to Islamism without an explicit statement by the data subject regarding her or his religion. The way in which this information is recorded is eventually crucial. In the former scenario, the record of an explicit self-assessment necessarily entails the categorisation of the collected data as special category; in the latter, such speciality is only implicit since religious believes can only be inferred after an in-depth scrutiny and by means of additional information. The same can be applied to veganism and vegetarianism and food-related health conditions.

The dishomogeneity in addressing data protection concerns can be due to multiple reasons, including:

(I) the wide time span in which data collection has been performed, which imply the shift from the Data Protection Directive to the GDPR in Member States legislation;
(II) the lack of a uniform guidance on key aspects regarding the essential data protection regime for food consumption data collection across the Member States;
(III) the derogation allowed by the GDPR for research and statistical purposes to be implemented by national laws (Article 89 GDPR). Moreover, EFSA 2019

protection. However, such wording does not offer us any possibility to review the extent to which the data controller has addressed data protection issues

[34] See, respectively, Dubuisson et al. (2017), para 2.6 and for Public Health et al. (2018), para 2.6

[35] Sette et al. (2011), p.923

[36] EFSA (2014), ch. 7

[37] Available at https://doi.org/10.5281/zenodo.1215993, accessed on 27/04/2022

decision on the processing of personal data[38] is not decisive in solving data protection concerns since it places risk assessment outside of its scope;[39]

(IV) the fact that data aggregation might induce data controllers in overlooking data protection measures even though data collection concerns information that is linked to identified individuals and data aggregation does not necessarily entail data anonymisation

(V) the variety of information-types at stake, which include native and derived/inferred sensitive information.

While the first three points (I, II and III) are peculiar to the domain at stake, the last two (IV and V) have been widely discussed by the literature on data protection. It is worth discussing the extent to which this discussion can affect the food safety domain and, in turn, if our scenario can bring additional contributions to this debate.

3.2.1 Quasi-Personal and Quasi-Sensitive Data

Inter alia, the previous section has identified two reasons for the dishomogeneity of the measures taken to mitigate issues related to the processing of food consumption data and some doubts as regards the applicable legal framework. To briefly restate the issues to be discussed, dietary data processing in food safety forces us to evaluate two aspects of their life cycle: on the one hand, data aggregation and its relationship with anonymisation; on the other hand, the generation of inferences and the legal framework that applies to them.

As regards the first set of issues, data aggregation is extremely frequent in food safety. Potential exposure to pathogens is measured by grouping individual food consumption data by age. Aggregation is considered one of the anonymisation techniques by the A29WP's opinion on Anonymisation.[40] Before discussing various

[38] Decision of the European Food Safety Authority of 19 June 2019 on internal rules concerning restrictions of certain rights of data subjects in relation to processing of personal data in the framework of the functioning of EFSA (2019) OJ L 272/154

[39] Article 2(2) of the Decision: [w]ithin the framework of the administrative functioning of EFSA, this Decision applies to the processing operations on personal data by EFSA for the purposes of conducting administrative inquiries, disciplinary proceedings, preliminary activities related to cases of potential irregularities reported to OLAF, processing whistleblowing cases, (formal and informal) procedures of harassment, processing internal and external complaints, conducting internal audits, investigations carried out by the Data Protection Officer in line with Article 45(2) of Regulation (EU) 2018/1725 and (IT) security investigations handled internally or with external involvement (e.g., CERT-EU)

[40] Opinion 05/2014 on Anonymisation Techniques, available at https://ec.europa.eu/justice/article-29/documentation/opinion-recommendation/files/2014/wp216_en.pdf, accessed on 27/04/2022

examples of anonymisation techniques, the opinion clearly details three risks which are essential to anonymisation:

- *Singling out*, which corresponds to the possibility to isolate some or all records which identify an individual in the dataset;
- *Linkability*, which is the ability to link, at least, two records concerning the same data subject or a group of data subjects
- *Inference*, which is the possibility to deduce, with significant probability, the value of an attribute from the values of a set of other attributes.

Considering the wide range of background information required by EFSA when building their datasets,[41] these risks are likely to materialise. Raw information remains available to EFSA and national data controllers. It is unclear whether the real name of the person providing the answer to food consumption survey is collected: an attribute displayed in the metadata ("PANSWER") seems to correspond to the identity of the surveyed individual, whereas the ORSUBCODE is used as a unique identifier. Either way, the collected background information seems sufficient at least to pose questions regarding the efficacy of grouping in effectively anonymising the data at stake.

While it is true that further releases of data (e.g., to commercial applicants willing to place regulated products in the market) may be anonymous if provided solely in aggregated forms, the availability of raw information (either in personal or pseudonymised forms) pose risks to fundamental rights and freedoms. When discussing "properly anonymised data", the A29WP clearly states that "even though data protection laws may no longer apply to this type of data, the use made of datasets anonymised and released for use by third parties may give rise to a loss of privacy. Special caution is required in handling anonymised information especially whenever such information is used (often in combination with other data) for taking decisions that produce effects (albeit indirectly) on individuals".

As it can be observed by the dishomogeneity in addressing data protection concerns, the legal classification of these data is cumbersome. When collected, they are linked to an identified individual; then, they are pseudonymised and processed in aggregated forms, in particular when shared with third parties. However, raw data still remain available at EFSA. Therefore, this environment is also characterised by the lack of a clear understanding of the nature of the data at stake. This seems consistent to the 'contextual' nature of the concept of personal data highlighted by a joint analysis by the European Data Protection Board (EDBP) and the European Data Protection Supervisor (EDPS)[42]

---

[41] Available at https://doi.org/10.5281/zenodo.1215993, accessed on 27/04/2022

[42] 'EDPB-EDPS Joint Opinion 03/2021 on the Proposal for a Regulation of the European Parliament and of the Council on European Data Governance (Data Governance Act)' (2021) p.15: "The EDPB and the EDPS underline that the distinction between categories of personal and non-personal data is difficult to apply in practice. Indeed, in practice, from a combination of non-personal data it is possible to infer or generate personal data, i.e. data relating to an identified or identifiable individual

Scholars[43] have discussed the limitations of the current definition of "personal data" under data protection law and jurisprudence. Their discussion has revolved around the transformative nature of data when certain data processing techniques (e.g., anonymisation, aggregation, re-identification by inference, and so on) are used. Taken together, this literature identifies how data can easily switch from one legal category to an other (e.g., personal to non-personal), which brings the consequence of the *ex-ante* uncertainty of the correct class and sub-optimal compliance.

Food consumption data processing is significantly affected by these limitations. Therefore, the critiques advanced by these authors suggest caution in the adoption of a straightforward classification of dietary intake information as 'personal data' and encourage an anti-formalistic approach to the question of legal categorisation for further discussion. To highlight the limitations in correct classification while maintaining an holistic approach to the presence of data *related to* individuals, food consumption data will be included within the notion of "*quasi*-personal data" for the purposes of this book.

Let us turn to the second set of issues, i.e., the possibilities offered by the use of food consumption data as proxy variables. In 2015, the A29WP "Letter to Paul Timmers" and its "Annex I on health data in apps and devices" identified a 'three-step approach' to assess whether or not personal information should processed as health data for the purposes of data protection law[44] when processed in the context of mobile applications:

1. The data are inherently/clearly medical data;
2. The data are raw sensor data that can be used in itself or in combination with other data to draw a conclusion about the actual health status or health risk of a person;
3. Conclusions are drawn about a person's health status or health risk (irrespective of whether these conclusions are accurate or inaccurate, legitimate or illegitimate, or otherwise adequate or inadequate).

The A29WP explicitly referred to a 'diet app' to explain the criteria of the 'three-step approach', as following: 'if a diet app only counts the calories as calculated from input provided by the data subject, and the information about the specific foods eaten would not be stored, it would be unlikely that any meaningful conclusions can be drawn with regard to the health of that person (unless the daily intake of calories is excessive in absolute terms). But if data from a diet app, or heart rate monitor or sleep diary app are combined with information provided by the data subject (directly

[43] See, *inter alia*, Purtova (2018), Malgieri (2016a), and Schwartz and Solove (2011)

[44] In the absence of the GDPR, the applicable legal regime was the Data Protection Directive (Directive 95/46/EC of the European Parliament and of the Council of 24 October 1995 on the protection of individuals with regard to the processing of personal data and on the free movement of such data OJ L 281/31). However, the two pieces of legislation do not different significantly as regards the definitions of data processing subject to stricter legal regimes due to the special nature of data undergoing processing

or indirectly, for example based on information collected from that person's social networking profile), conclusions (whether accurate or inaccurate) may be drawn about that person's health condition, such as medical risk or diabetics'.

However, in its attempt of clarifying the regime of proxy data and dietary intake information indirectly, the A29WP also raised new questions and doubts. How 'excessive' should be the amount of calories for it to be considered health data? What happens if user-provided information is occasionally excessive, thus being only a portion of his/her data subsumed under the category of health data? Is explicit consent required every time the user eats to much?

Considering the explanatory nature of the example provided by A29WP and the doubts raised by a literal interpretation of what has been stated about diet apps, we will not consider this statement as decisive to the inquiry of food consumption data legal regime. However, it appears beyond doubt that food diaries were subsumed into the category of 'fitness apps' in the context of mobile health. However, food consumption data are processed in several ways other than food diaries, including in the context of public risk assessment.

While the implications of the processing of background information (age, sex, marital status, region, rural/urban area, size of the household, household income, employment status)[45] are widely covered by the literature on privacy and data protection, studies on the processing of food consumption data and dietary patterns have been focusing on their proximity to the disclosure of health data. For instance, the discussion on *quasi*-health data[46] and the context of mobile healthcare have occasionally included a new category that is placed in-between the legal classes of "regular" personal data and special categories of data.

However, despite the relevance of mobile health apps and the likelihood of health data discovery by data processing, our domain is far more intricate than a smartphone-mediated data gathering methodology that might reveal sensitive attributes. In fact, food consumption data can be used to generate inferences other than the status of health, including philosophical (veganism, vegetarianism) and religious believes (e.g., due to fasting). Therefore, the proximity to health data would offer a limited perspective: while it might be appropriate to other information-types which only can relate to the state of health (e.g., footsteps, heartbeat), food consumption information is close to multiple special categories of data.

As with the matter of approaching dietary information as *quasi*-personal data, *quasi*-sensitive is a convenient perspective to analyse food consumption information. The rationale underlying this choice is the necessity of preserving consistency with existing legal categories while underlying the high inferential power of dietary information, in particular towards special categories of data.

The ongoing debate about *quasi*-personal and *quasi*-sensitive categories deserves attention. It could be enriched by discussing food consumption data as a peculiar

45 See https://doi.org/10.5281/zenodo.1215993, accessed on 27/04/2022

46 Malgieri and Comandé (2017)

information-type. The next section attempts to provide food for thought on this research trend and discuss its relation to dietary data.

3.2.2 Food Consumption and Limitations in the Notion of Personal Data

In the previous section, *quasi*-categories were introduced and a general assumption about their appropriateness for the context of food consumption data was made. It is now time to discuss the origin of these categories and how the ongoing debate can make use of food consumption data. To summarise the two perspectives under discussion, the sets of issues regard the classifications of "personal/non-personal" and "regular/sensitive" data. The latter necessarily entails that the data undergone processing are classified as personal under data protection law.

The aforementioned scholars found limitations in legal categories used in data protection laws. Purtova argues that the interpretation of 'personal data' under data protection law is broad enough to include almost all information available.[47] This is due to a combination of phenomena such as the 'datafication' of human behaviour and the ongoing shift towards 'smart' and connected environments. As a result—she argues—"it is safer to assume that any information is likely to relate to a person". Ohm reached similar conclusions when observing, *inter alia*, that 'easy reidentification makes laws like the EU Data Protection Directive boundless and overbroad' as almost every piece of information can be related back to individuals.[48] Purtova also observed a certain consistency in broad interpretations of the concept of personal data in the case law of the ECJ and scholars' opinions.[49] Her proposal consists of abandoning "the concept of personal data as a cornerstone of data protection altogether and seek remedies for 'information-induced harms'—understood broadly as any individual or public negative consequences of information processing—without a sentimental attachment to this familiar proxy".[50]

Moreover, Malgieri found that classifications related to certain subject matters (e.g., health, genetic, financial, commercial) are anachronistic and proposed a relationship-based taxonomy of personal data characterised by three degrees of proximity to the data subjects.[51] His proposal revolves around the concept of relation in the definition of personal data, i.e. information *relating to* an identified

---

[47] Purtova (2018)

[48] Ohm (2009)

[49] Purtova (2018)

[50] Alternatives such as the narrowing the broad interpretation of personal data or lowering compliance burdens for data controllers are also mentioned, yet discarded. Notably, Schwartz and Solove (2011), p.1866 note that "abandoning PII [*ed.* see below] is problematic, however, because the concept serves a crucial function: it establishes the boundaries of privacy regulation"

[51] Malgieri (2016b)

or identifiable natural person. Consequently, the same author, as anticipated above, proposed the *quasi*-health category, once again relying on a fictional in-between class that measures the proximity to one sensitive attribute, namely health conditions.

In the US legal system, scholars[52] have noted that the origin of the concept of PII, or Personal Identifiable Information,[53] can be traced back to the privacy as right of 'personality' in the foundational 1890 article entitled 'The Right to Privacy' by Warren and Brandeis.[54] Identifiability is a necessary precondition of possible privacy invasions. More importantly Schwartz and Solove[55] describe 'the anonymity myth', which "stems from a mistaken conflation between momentary anonymity with actual untraceability", when referring to allegedly anonymous browsing information. They support their argument with a significant number of examples, all showing that the personal/non-personal nature of information depends on the context of processing and on the technology used in the data analysis (p. 1848). In the same legal system, Ohm[56] expanded the scope of the issues pertaining to the personal/non-personal categorisation of data to the regular/sensitive nature of the concerned information.

If data undergone processing are classified as personal data, then other issues related to their sensitive nature can arise. Malgieri and Comandé[57] discussed the relationship between regular and health data, also considering dietary intake information. Their study proposes a 'sensitive-by-distance' approach: the proximity of personal information to health data is measured by its 'intrinsic sensitiveness' and the 'computational distance' that separate the raw piece of information from revealing the status of health of an individual, in conformity with the definition given by Recital 35 of the GDPR. Then, according to this measure, the applicable legal category is identified. The computational distance includes by definition the possibility of deriving and inferring sensitive data also by means of other information. While the presence of such a dynamic perspective seems appropriate to the nature of food consumption data processing, this proposal seems to neglect that, within the cited information-types, food consumption data are not close to health information only. Instead, they can reveal other sensitive traits of individuals, such as religious and philosophical believes. Nonetheless, their study correctly highlights limitations in data protection law in addressing the cross-category nature of inferences.

The issue underlying the legal categorisation could be the necessity of identifying a precise material scope for the applicability of general or specific data protection

---

[52] See Schwartz and Solove (2011)

[53] Purtova (2018) argues that personally identifiable information are a functional equivalent of personal data in the USA

[54] Brandeis and Warren (1890)

[55] Schwartz and Solove (2011)

[56] Ohm (2009)

[57] Malgieri and Comandé (2017)

rules. On the one hand, the boundary between personal and non-personal data is necessary to require compliance for certain data processing activities. On the other hand, drawing a line between regular and special categories of data is required because certain information-types shall be subject to stricter rules due to the increased risks for the fundamental rights and freedoms linked to their processing.[58]

Data protection laws thus apply a strict binary classification which contradicts the outcome of certain data processing activities, including in the field of food consumption data. Re-identification and inferences pose questions to the resilience of binary legal categories. In turn, these difficulties jeopardise, as observed, the level of protection of food consumption data and bring sub-optimal compliance.

Solutions are proposed. For instance, as regards the issue of re-identifiability, scholars[59] have argued that "a better solution would be, first, to view the identifiability of data as a continuum as opposed to the current dichotomy". Malgieri and Comandè's[60] approach seems to share the same continuum perspective, even though it is functional to identify the appropriate legal category.

As with other information-types, food consumption data are hard to classify within the existing legal categories and continuously cross them. In comparison to other personal data, dietary information processing is characterised by unclear data aggregation/anonymisation methods and high inferential power, especially when combined with other personal data. *Quasi*-categories are placed outside the existing legal framework and shall be considered proposals for further discussion. Their merit, however, is to highlight limitations in data protection law which are reflected on food consumption data processing. Likewise, the collective dimension of group privacy casts light on other issues pertaining to data aggregation.

3.2.3 Group Privacy and Dietary Preferences

Another anti-dogmatic perspective, placed outside the current legal framework yet in relation to existing norms is the one of 'group privacy'.[61] It has been noted that data aggregation and group-level analyses are common practices in food safety risk assessment. Therefore, their implications can also be discussed in the light of their collective and social impact.

A preliminary issue to be discussed in the conceptualisation of group privacy is whether or not groups constitute a mere 'sum of individuals' or they exist as a separate and independent entity. Scholars[62] first recognised that groups are usually dynamic entities, hence the difficulty in identifying stable properties (*rectius*,

[58] See Recital 51 of the GDPR

[59] Tene and Polonetsky (2012), p.259

[60] Malgieri and Comandé (2017)

[61] Taylor et al. (2016a)

[62] See Taylor et al. (2016b)

predicates) that fix them. This might imply that, without stable properties, the only fixed point would hypothetically be the individual, hence the conceptualisation of groups as 'sum of individuals'. In their words, this constitutes the "individual-based, entity-first approach".

However, this line of reasoning neglects that it is the choice of the predicate (e.g., "vegan" or "vegetarian") that determines who belongs to the group. Then, as the authors note, "it is the same digital technologies used to create a group by selecting some properties rather than others (e.g., "Muslim" instead of "Christian") that can also infringe its privacy. Technologies actually determine groups, through their clustering and typification". In their words, this is the "property-based, predicate-first approach". Authors conclude that "if there is such thing as group privacy it is to be analysed as an emergent property, over and above the collection of the privacies of the constituting members".

Group privacy seems an intriguing perspective in the analysis of food consumption data processing. For instance, limitations in the applicable law highlighted in the previous subsections can be mitigated when discussing informational privacy as regards aggregated data that refer to individuals that are clustered according to dietary patterns.

Questions arise when applying the same reasoning to food consumption data. What are 'groups' in this scenario? Dietary preferences can be aggregated in several ways. Usually, aggregation is performed by age of consumption.[63] Nonetheless, it might be the case that other groups can be designed[64] by machines autonomously.

To investigate such autonomy, I will introduce two categories. Inferences from individual data points can be deemed to be *intended* or *unintended*. In the former case, an epistemic agent (e.g., a data scientist) looks for specific correlations that equal to a pre-defined category. For instance, he or she could be willing to separate "male" from "female" consumers. Thus, he or she could look for correlations and patterns that allow him or her to aggregate individuals according to such pre-defined categories. Finally, he or she instructs the machine to design the group accordingly. As we are still in the realm of inferences, the belonging to a certain category is *likely* or expressed in terms of probability. Instead, the categorisation in the case of *unintended* inferences do not constitute the goal of data analysis. For instance, if a clustering algorithm is deployed to group users according to their preferences—for instance, to identify patterns useful for risk assessment[65]—it could aggregate individuals according to similarities in food consumption patterns without any *a priori* instruction and definition of the categories that have be discovered. When the output is visualised by the data scientist to evaluate the efficiency of the algorithm in grouping users, he or she is unintendedly visualising groups of vegetarians, Muslims or coeliac, possibly even without noticing it.

---

[63] EFSA (2019)

[64] Floridi (2017) argues, consistently with the "property-based, predicate-first approach" approach that groups are "neither discovered nor invented, but designed by the level of abstraction (LoA) at which a specific analysis of a social system is developed"

[65] Hearty and Gibney (2013)

The main issue with unintended inferences is that the attribution of a given label to an identified group does not occur even *ex-post* since the categorisation is somewhat implicit. This might lead to subtle forms of discrimination that fall outside the scope of anti-discrimination law, yet based on sensitive attributes such as health conditions, religious or philosophical believes. What makes inferences drawn on food preferences peculiar is that inferred categories correspond to properties that fall within the special categories of data identified by data protection law.[66] This trend is not exclusive to food safety: other scholars[67] have found similar trends in other domains.

Considering the aforementioned limitations in data protection law due to its categories (personal/non-personal and regular/special categories) and the challenges raised by food consumption data processing, a group-level discussion can be a viable way to understand what can be done to mitigate issues related to their analysis, including security breaches and discrimination.

3.3 "You Are What You Eat": Food and Personal Identity

If a group-level discussion capable of understanding the collective and social implications of food consumption data processing is a viable way, then it might be the case to verify how the scholarship on group privacy can contribute to our discussion. This section first briefly introduces privacy theories, then contextualises the datafication of food preferences and their relations with personal identity, before broadening the scope of analysis to include governance and policy options.

Floridi[68] argues in favour of a quadripartition of the concept of privacy:[69] *physical* privacy protects the individual from tangible interferences against her or his body; *mental* privacy prevents others to manipulate one's mind and protects the psychological dimension of an individual; *decisional* privacy protects individuals from interferences in decision-making processes; *informational* privacy aims to protect individuals from interferences that related to facts that shall be not be disclosed. The concept of informational privacy is broader than data protection, as it does not only cover "information related to an identified or identifiable individual" but also "facts", which might be placed outside the scope of data protection law.[70] In practice, however, this difference tends to disappear and this is particularly true in

[66] Art. 9 of the GDPR, Art. 10 of Regulation 2018/1725

[67] Durante (2019), p.259

[68] Floridi (1999) and Floridi (2013)

[69] Other approaches are those discussed in Tavani (2008), footnote 12–13

[70] As pointed out by one reviewer, the two concepts could also be treated separately and independently. However, it has to be noted that the aforementioned literature tends to treat them jointly and focuses on differences and overlaps. Consistently with this approach, I have opted for a joint discussion

the context at stake. As argued, the human behaviour of eating ("facts") is recorded and associated to a given individual ("data"). However, in light of our technical contextualisation, it is first necessary to verify what facets of privacy can be applied in our domain. Privacy rationales discussed below are taken from the reviews made by Tamò-Larrieux.[71]

The Restricted Access/Limited Control (RALC)[72] and related theories understands privacy as a way to exclude others from a private 'zone' or 'sphere'.[73] These approaches are rooted in the well-known 'right to be let alone'[74] that was originally discussed as a response to intrusions in the private sphere of an individual. Today, these theories conceptualise informational privacy as limited access to one's personal information.[75] Similarly, the ownership-based interpretation of privacy argues that individuals claim property over information regarding them,[76] thus having recognised an *ius alios escludendi*.

Then, the contextual integrity framework[77] suggests that the nature of the processed data is shaped by the informational norms and the context concerning the flow of information. Therefore, whenever the processing of certain personal data is inappropriate for the context in which this flow of information takes place, some kind of protection should be given.

The risk-based (or reductionist) interpretation of informational privacy understands data processing as a hazardous activity and focuses on the technical and legal remedies to mitigate the risks connected to it. Among many scholars, De Hert and Gutwirth[78] noted this approach is pursued by the European Courts of Human Rights. Similarly, Lynskey[79] provided examples of anti-discrimination and anti-identity theft legal provisions aimed at minimising data processing risks. The Data protection Impact Assessment (DPIA) prescribed by Art. 35 of the GDPR is mandatory if it 'is likely to result in a high risk to the rights and freedoms of natural persons', thus implicitly endorsing a risk-based interpretation of privacy when additional burdens are placed on data controllers due to potential harms resulting from the processing of data.

Finally, Floridi's ontological interpretation[80] (or theory, as he refers in a later study[81]) of informational privacy understands personal data as a constitutive part of one's personal identity and promotes data protection as a right connected to

[71] See the work by Tamò-Larrieux (2018)

[72] Moor (1991), Moor (1997), and Tavani and Moor (2001)

[73] See also ownership models in Sect. 2.1.1

[74] Brandeis and Warren (1890)

[75] Moor (1997)

[76] Stepanov (2020)

[77] Nissenbaum (2020)

[78] De Hert and Gutwirth (2009)

[79] Lynskey (2014)

[80] Floridi (2005, 2013)

[81] Floridi (2006)

human dignity.[82] The author notes that "the ontological interpretation stresses that informational privacy is also a matter of construction of one's own informational identity". Some timely critiques towards the lack of a clear boundary separating loss of privacy from violations of privacy have been proposed.[83] However, this approach seems to be echoed by some provisions on the GDPR that allow the data subject to correct or erase her or his data to mirror her or his personal identity (e.g., art. 16, art. 17).

The next section will discuss why such ontological approach might be appropriate to frame data-related privacy issues in food safety data processing. This does not entail, however, that the ontological theory shall be considered the best-working approach for every kind of data processing. Our claim shall be solely confined to the domain under scrutiny. Therefore, also the following limitations of other theories should not be considered as absolute limits, but relative to food consumption data and related practices.

Considering RALC-wise theories and taking into account food consumption data processing practices and inferences that can be drawn from them, 'zones' and 'spheres' constantly change their area and volume. Food consumption data flow across Europe and worldwide, through various entities, since such information sharing is crucial for a timely and accurate risk assessment: while data subjects would be eager to share their data to improve the safety of foodstuffs, they possibly would not let industries in certain fields—including pesticides or GMOs—for personal reasons, such as their ethical thoughts. However, data processing practices at stake suggest that the law does not contemplate this option for the data subjects, regardless of their opinion.

The limitation of the contextual integrity framework lies on the presence of a vertical shift of the flow of information. Nissenbaum describes a flow of information ($\rightarrow$) in terms of transmission, communication, transfer, distribution, and dissemination of data from one entity (A) to other(s) (B). In the inferential scenario at stake, such 'horizontal' flow ($A \rightarrow B$) is not appropriate, since the presence of inferences creates a 'vertical' dimension in which every actor put efforts in discovering new information by transferring data to a machine (M) that belongs to the same actor (A), in a flow that has no horizontal dimension due to the absence of other parties ($A \uparrow M$). At the end of the analysis, data is then returned to the actor ($M \downarrow A$). One could argue that, even the 'vertical' flows of information ($A \updownarrow M$) are governed by informational norms. However, while this can be true for the $A \uparrow M$ flow, in which the controller pre-determines the sensitive/non-sensitive nature of data and acts compliantly, the normative context of the return flow $M \downarrow A$ is shaped by attributes (sensitive/non-sensitive) discovered by algorithms and possibly unknown to the data analyst. These machine-determined informational norms may

[82] See also European Data Protection Supervisor's Opinion 4/2015 "Towards a New Digital Ethics Data, Dignity and Technology"

[83] Tavani (2008)

be inappropriate for the real nature of the transferred information, i.e. sensitive information related to an individual derived from food consumption data.

One limitation affects the reductionist interpretation of privacy. The cost-benefit analysis assessing whether protection should be given to certain data including dietary intake information may result in accepting the risks linked to the possibility of drawing sensitive inferences for individual and social welfare. As we noted, information sharing is both convenient and pursued in the context of food safety. As the sense of belonging to an ethical movement is crucial in maintaining a vegan lifestyle, individuals may be willing to share their preferences also within a digital environment. Environmental campaigns targeting vegans or vegetarians can ultimately bring positive effects on the society by raising attention on topics such as animal welfare and food waste reduction. Therefore, the balancing of conflicting rights can be solved in favour of interests other than food consumption data privacy by referring to a generic 'risk mitigation' approach.

3.3.1 The Datafication of Food Preferences

Before entering the discussion about an identity-constituting approach to food consumption data processing, it is briefly necessary what this practice entails. Purtova noted that "everything in this environment—weather, waste water, exam scripts—is being increasingly 'datafied'".[84] Food preferences are not exceptional. Then, what are data? How do they relate to food preferences?

Floridi argues that a Datum is an ontological "lack of uniformity"[85] in the world. He discusses a diaphoric interpretation, based on differences. Divergences can be observed in the real world (Data as diaphora *de re*), in the perceptions of signals, e.g., Morse code lines and dots (Data as diaphora *de signo*) and among symbols, for examples two letters from an alphabet (Data as diaphora *de dicto*). It follows that different symbols encode different signals that reveals anomalies around us.

Therefore, datafication can be defined the recording of *differences* among real-life phenomena. While this process can occur also in paper, immaterial forms are usually the focus of studies on datafication. In Floridi's interpretation, the concept of data includes: primary data (i.e. array of numbers stored in a given support), secondary data (i.e. the absence of data, like the silence in a communication), metadata (i.e. indications related to primary or secondary data), operational data (i.e. data about the functioning of an information system) and derivative data (i.e. data that can be generated from some other data where the latter is used as the source of patterns and inferences). Let us focus on these aspects as well.

What elements are kept into account when food preferences are datafied? Individual food-related behaviour is the first object of datafication. Depending to

[84] Purtova (2018)
[85] Floridi (2013), p.85

the selected methodology of data collection (discussed above), the recording may be self-assessed or recorded by some other agent. Then, a dietary pattern is a form of *derivative* data that, by aggregating individual consumptions, might lead to the identification of regularities in diet of a given individual. From patterns, *inferred* data can be generated by associating certain habits to a given label (e.g., Mediterranean diet, low carb, keto, etc.). In some instances, this inferential process can reveal attributes that identify other traits of the individual, such as her or his health status, religious or philosophical believes.

The lack of recorded consumption becomes relevant in food patterns. As we stated above, it is the lack of meat in one's diet that reveals adherence to vegetarian diet. Similarly, the lack of alcohol or pork can show someone's adherence to Islamism. These absences can be used as predictor to explicitly infer other information about individuals. Moreover, as argued above, such inferential process is made possible by the abundance of background information collected for risk assessment processes and the inference can also be drawn unwillingly by the (data) scientist when grouping or aggregating data.

Therefore, the datafication of food preferences shall not be considered as an isolated occurrence that is limited to the collection stages. In food safety risk assessment, data is *collected*, *derived*, and *inferred*. The transformative nature of food consumption data processing has to be understood as the 'computational power'[86] to reshape food-related behaviours of individuals.

Such power entails at least two sets of risks briefly discussed in the previous sections. First, we noticed in Sect. 3.2 that compliance to data protection law is jeopardised at the collection level when national authorities gather dietary information and background data. A tentative explanation of the sub-optimal compliance is a certain degree of opacity in the nature of food consumption data between personal/non-personal and regular/sensitive. Then, the combination of aggregation and inferences could lead to group-level discrimination: if scientific results related to certain groups, including those created unwillingly by the machine, are neglected, risks of discrimination become possible. As argued above, while data protection can mitigate these issues, a unified rationale inspiring or justifying measure is still missing. The next section explores how an identity-constituting approach to informational privacy can serve as a unified rationale for risk mitigation measures.

3.3.2 From Food Preferences to Personal Identity

Let us briefly restate why a focus on personal identity is necessary. So far, this section has discussed food consumption data processing practices. First, a technological review has shown how this personal information is collected for food safety purposes and how such collection might also occur with novel and emerging

[86] Durante (2019)

methods. Then, the nature of intake information as *proxy* has raised concerns about sub-optimal compliance shown by food safety authorities and the possibility of group discrimination. While these issues are presented and discussed separately, they are inextricably linked to the *datafication* of food preferences.

The first set of issues was justified by the nature of the data at stake and their being cross-categorical: since they are placed in-between existing legal categories in data protection law, full compliance can be hard to be achieved and maintained. The second set of issues is linked to how grouping takes place. Since food-related information is used to categorise individuals and predict risks for their health, it may be the case that scientific evaluations do not keep into account specific groups. In some cases, these groups mirror existing categories which are canonical grounds of discrimination (e.g., religious believes or ethnic origin), emerging concerns (e.g., veganism or vegetarianism). Let us explore avenues to provide a unified rationale capable of justifying *why* these issues shall be solved.

Personal identify has received attention in philosophy—even before Socrates[87]—bioethics[88] and legal informatics, especially in the context of online behaviour and representation[89] and digital identity.[90] A conceptual separation between the *public* identity, i.e. the one recognised or attributed by the State to exercise its functions, and the *substantial* identity, i.e. the one that any individual builds and shows to the community, has also been made.[91] Only the latter concerns this study.

Recently, Martoni and Pattaro[92] have proposed the concept of *self-representative personal identity* to define the image made up by someone and expressed in public to the rest of one's community. This relational dimension entails that one's personal identity does not consist of one's self-image, nor it amounts to the history of an individual.[93] Moreover, the social component is a further justification of the right to privacy as the right to protect ourselves when interacting with the public.[94] This will be the definition of identity adopted in this subsection.

Food preferences, whether they are entirely subjective or related to other physical conditions (allergies, intolerances, etc.), contribute to make each person a unique individual. Studies in neuropsychology suggests that children progressively align their food preferences to adults only by growing and developing a more complete body.[95] Moreover, let us think to Brillat-Savarin's most famous quote "Tell me what you eat, and I will tell you what you are"[96] or Feuerbach's: "Man is what he eats".[97]

---

87 Martoni and Pattaro (2021)

88 Pino (2010)

89 Martoni and Palmirani (2015) and Finocchiaro (2010)

90 Palmirani and Martoni (2006)

91 Zencovich (1993), p.294

92 Martoni and Pattaro (2021), p.69

93 Finocchiaro (2010), p.725

94 Martoni and Pattaro (2021), p.78

95 Rozin et al. (1986)

96 Brillat-Savarin (1841)

Though in broad terms, these examples show some degree of connection between food preferences and personal identity.

The relationship between what we eat and our identity is also true as regards the belonging to certain social groups. We occasionally refer to other groups and people using the suffix *-eaters* or expressions such as "Frogs" for French, "Krauts" for Germans, "Macaronis" for Italians.[98] The same holds true for particular cuisines when they are associated to certain nationalities (like the Italian or the French one) in a manner that makes the cuisine peculiar to a national cultural heritage. The adherence to a particular diet—like veganism—can express one's belonging to a cultural movement[99] or a philosophical belief. A British employment tribunal[100] found out that veganism meets the five criteria to qualify as a philosophical convictions: (1) it must be genuinely held; (2) iIt must be a belief and not an opinion or view point based on the present state of information available; (3) it must be a belief as to a weighty and substantial aspect of human life and behaviour; (4) it must attain a certain level of cogency, seriousness, cohesion and importance; and (5) it must be worthy of respect in a democratic society, not incompatible with human dignity and not conflict with the fundamental rights of others. Hence, it cannot be considered "taste".

Religions make a wide use of food-related prescriptions (including fasting) to reinforce the belief: Catholic Canon Law prescribes the abstinence from meat on every Friday, Ash Wednesday and Good Friday for everyone between 14 and 60 years of age; during the month of Ramadan in the Islamic calendar, fasting is recommended from dawn to sunset (daylight hours); in the Kashrut dietary laws, consuming animals listed in the 613 commandments is strictly prohibited. Despite the paucity of studies on this topic, political opinions are assumed to play a role in determining food preferences. Inferences drawn from alcohol consumption proved association between beer and spirits drinking habits and liberal ideology, holding economic, demographics and geographic differences constant.[101] In a different study, it has been proposed that "left-wing" food habits are sensitive towards environmentalism, organic food and *farm-to-table* markets, whereas the "right-wing" ones tend to prefer frozen food, massive portions and energy drinks.[102] Research has also focused on predicting overweight and diabetes (plus, incidentally, political opinions) rates from a large corpus of Tweets containing food-related hashtags.[103] Although this example is not directly referred to recorded food consumption, it confirms that food-related behaviours can be used as predictors for personal characteristics beyond food consumption.

---

[97] Cherno (1963)

[98] Fischler (1988)

[99] Cherry (2006)

[100] Case Casamitjana v The League Against Cruel Sports [2020] UKET 3331129/2018

[101] Yakovlev and Guessford (2013)

[102] Sasahara (2019)

[103] Fried et al. (2014)

Let us discuss some findings from the joint analysis of the discourses on personal identity and food consumption datafication. Floridi's theory of privacy revolves around personal identity. The identity-constituting interpretation stresses that privacy is also a matter of construction of an individual's own identity.[104] In our reconstruction, we adopted a concept of identity as *self-representation* before the society. Contextualised in a datafied scenario, *self-representation* would necessarily imply *self-determination* against manipulation. However, this cannot be the case for the data processing practices at stake—in particular, anonymisation and inferences—that consistently limit the applicability of data protection rights.

Despite these limitations, acknowledging the proximity of food consumption data to personal identity is necessary to mitigate risks of poor compliance and group discrimination highlighted in the previous subsections. Such acknowledgement entails the necessity of adopting mitigating measures justified by the fact that, despite anonymised or inferred, the information undergoing processing is still related to human beings and expresses their peculiar traits.

Let us also consider the advantages of an identity-constituting approach in comparison to other theories on privacy rationales. Firstly, it does not create property-like personal rights over personal data and inferences drawn on them, thus avoiding potential conflicts between different owners. However, it still grants protection over personal data without diminishing the correlation between an individual and his or her data, as 'his' and 'her' are understood as integral parts of data subject's individuality rather than his or her property. Secondly, it synergises with the contextual integrity framework since every actor involved in data processing recognises that informational norms governing the 'vertical' flow should keep into account the close connection between personal/social identity and dietary intake information. Therefore, the issue related to the machine-determined context introduced in the previous section can be mitigated by the human-centred consideration of the information at stake. Hence, the M $\downarrow$ A flow can be deemed to have certain attributes by acknowledging the strict connection between the information involved in the process (i.e. inferences drawn upon food consumption data) and personal identity. Thirdly, this awareness also mitigates the risks related to the irrebuttable presumption of primacy of certain fundamental rights over others, possibly grounded on utilitarian perspectives, as in the case of the preponderance of the collective right to have a safe food over the individual rights to informational privacy. By conceptualising food consumption data as a proxy variable for attributes that have a strict connection with our identity, such balance of interests would no longer consist of a mere cost-benefit analysis, but the outcome of a thoroughly thinking process that involves human dignity and fundamental rights. Lastly, as regards unintended inferences, this approach could be determinant in reinforcing the awareness towards potentially discriminatory outcomes, thus steering the reductionist approach towards a more identity-constituting consideration of the risks at stake. A corollary of this approach is its potential application to the expectations

[104] Floridi (2017), p.96

of data controllers processing dietary intake information. By acknowledging the correlation between personal identity and food consumption, their foresight of sensitive inferences is likely to be, at least, more reasoned.

To summarise, I would suggest that the processing of raw food consumption information entails the observation and the recording of characteristic traits of individuals and groups that regard their personal identity. An identity-constituting approach is also more functional in comparison with other approaches when discussing the nature of the data at stake. This finding justifies measures aiming at promoting compliance and protecting individuals from discrimination. This subsection has described *why* measures shall be taken. The next one explores *what* can be done under a *de iure condendo* perspective.

3.3.3 The Next Steps in Dietary Intake Data Protection

The previous sections have identified some degree of connection between personal identity, food consumption datafication and identity-based approaches to informational privacy. Taken together, these findings recommend the acknowledgement of the role of personal identity as the justification of risk-mitigation measures even when data protection laws are not applicable. *What* can be done on the basis of this rationale?

As regards individuals, an identity-constituting theory of privacy can advantage data subjects for what concerns the protection of her or his datafied dietary behaviours by raising awareness towards data-related issues. Not only does this consideration apply in the context of food safety, but it is also useful in other datafication scenarios (e.g., mobile apps) when food consumption data are at stake and let alone other personal data that are processed in this context.

For instance, data analysis shall be understood as a form of manipulation of one's identity, with inferences being the discovery of new traits of an individual, even when she or he is no longer identifiable. From an identity-constituting perspective, grouping entails the attribution of a new predicate to an individual. Awareness towards grouping and aggregation can suggest measures aimed at exploring such manipulation and minimise connected risks, for instance in the context of the DPIA. At the same time, the increased relevance of this informational component in food safety can stress the necessity of deploying security measures suitable of protecting this information against unlawful exploitation.

Following the discourse on group privacy, further insights on group privacy positions are those provided by Floridi[105] and they are necessary to understand what can be done to protect groups. He identifies:

- a sceptics position, arguing that rights are properties that qualify only members of a group, not the group itself;
- a moderate position, arguing that rights might be attributed to groups only if their members enjoy these rights;
- a strong position, arguing that here are some kinds of rights that belong only to a group as a group.

Floridi[106] argues that "looking at the nature of an individual person or a group as being constituted by that individual's information enables one to understand the right to privacy as a right to immunity from unknown, undesired, or unintentional changes in one's own identity as an informational entity, both actively and passively. Actively, because collecting, storing, reproducing, manipulating etc. Alice's or her family's information amounts now to stages in cloning and fixing (profiling) their identities. Passively, because breaching Alice's or her family's privacy may now consist in forcing the individual or her group to acquire unwanted information, thus altering their nature as informational entities without consent".

Then, taking together privacy rationales and group privacy positions, Floridi's comprehensive analysis suggests that an identity-constituting theory of privacy can be applicable to groups according to a strong interpretation of group rights. This is particularly convenient in our scenario because individuals' rights over their data can be limited or not applicable. Therefore, although data subjects cannot enjoy their rights, groups to which they belong can be kept into account to mitigate risks of group discrimination and, indirectly, protect the individuals. Risk assessment is based on data that might mirror a religion or a philosophical belief: acknowledging the underlying or latent existence of these groups can be a step towards neutral, unbiased, and non-discriminatory results. However, this interpretation calls for a particular conceptualisation of the notion of "fairness" that I will describe later on. While acknowledging the existence of collective rights to informational privacy might minimise certain risks associated to data clustering, it is crucial to remark that group privacy is currently placed outside the current legislative framework, thus self-regulation being the most viable way for implementation.

A plethora of options is available to put this conceptual framework in practice. These alternatives include code of conducts and self-regulation mechanisms, but also authoritative interpretations, such as the European Data Protection Board. Implementations shall also keep into account other emerging concerns discussed in the following Chapter and other principles whose alignment is necessary to provide a robust governance framework.

---

105 Floridi (2017)

106 Floridi (2017), p.94

3.4 Chapter Synopsis

Food consumption data are not exceptional (in the sense they do not constitute exceptions to the general data protection framework) but pose relevant question to data privacy scholars discussed in this chapter.

First, a preliminary operation was to identify data collection and analysis trends. This was deemed necessary to identify data-related practices in the area of food consumption information. Our analysis has focused both on food safety data collection and 'fitness apps' that gather food consumption information. This is not a trivial exercise, because data protection authorities and scholars have referred to smartphone apps when discussing this information-type.

While some of their conclusions are applicable to the domain of food safety, the context at stake suggests a careful analysis on the practices adopted in data processing. In particular: (a) data aggregation limits the applicability of data protection laws, but raw information remains available, and (b) other data are provided alongside food consumption data.

As a consequence, individual dietary data are placed in-between legal categories: on the one hand, they are *quasi*-personal for being collected as personal, then aggregated for further analysis, while still being linkable to individuals, also because of the background information that are provided alongside consumption; on the other hand, while they are classified as "regular" personal data, they allow, in combination with other information, inferences capable of discovering sensitive traits of individuals (religious believes, health status, and so on). Therefore they are also *quasi*-sensitive.

Data protection law does not contemplate these *quasi*- classifications, thus an anti-formalistic approach being necessary. Issues raised by the possibility of exploiting food consumption data for purposes other than risk assessment and/or discriminating on the basis of inferences drawn from them suggest that measures shall be taken to mitigate these risks. Why?

Our reply was an identity-based approach as a unified rationale. Following Floridi's interpretation or theory of informational privacy, an identity-constituting approach was chosen as the rationale inspiring risk-mitigation measures. Dietary preferences are close to our identity ('you are what you eat') and therefore food consumption data would eventually be protected—even when they fall outside data protection law—due to their link with personal identity, which correspond to a rationale inspiring privacy measures. Also, this approach mitigates some weaknesses of other privacy rationales.

Some implications and approaches are also discussed. Risk minimisation measures shall be aimed to promote security, fairness and non-discrimination against risks that can emerge from the processing of food consumption data. A group-level perspective seems necessary to frame fairness and discrimination issue.

References

Abrams M (2014) The origins of personal data and its implications for governance. http://informationaccountability.org/wp-content/uploads/Data-Origins-Abrams.pdf

Ambrus Á, Horváth Z, Farkas Z, Dorogházi E, Cseh J, Petrova S, Dimitrov P, et al (2013) Pilot study in the view of a Pan-European dietary survey – adolescents, adults and elderly. EFSA Support Publ 10(11):508E. https://doi.org/10.2903/sp.efsa.2013.EN-508. eprint: https://efsa.onlinelibrary.wiley.com/doi/pdf/10.2903/sp.efsa.2013.EN-508

Archer E, Pavela G, Lavie CJ (2015) The inadmissibility of what we eat in America and NHANES dietary data in nutrition and obesity research and the scientific formulation of national dietary guidelines. Mayo Clinic Proc 90:911–926. 7. Elsevier

Batis C, Mendez MA, Gordon-Larsen P, Sotres-Alvarez D, Adair L, Popkin B (2016) Using both principal component analysis and reduced rank regression to study dietary patterns and diabetes in Chinese adults. Publ Health Nutrit 19(2):195–203

Brandeis L, Warren S (1890) The right to privacy. Harvard Law Rev 4(5):193–220

Brillat-Savarin, JA (1841) Physiologie du goût. Charpentier

Cherno M (1963) Feuerbach's "Man is what He Eats": a rectification. J History Ideas 24:397–406

Cherry E (2006) Veganism as a cultural movement: a relational approach. Soc Movement Stud 5(2):155–170

De Hert P, Gutwirth S (2009) Data protection in the case law of Strasbourg and Luxemburg: constitutionalisation in action. In: Reinventing data protection?. Springer, Berlin, pp 3–44

Dubuisson C, Carrillo S, Dufour A, Havard S, Pinard P, Volatier J-L (2017) The French dietary survey on the general population (INCA3). EFSA Support Publ 14(12), 1351E

Durante M (2019) Potere computazionale: L?impatto delle ICT su diritto, società, sapere. Mimesis

EFSA (2011) Activities, processes and quality assurance elements on data collection programmes with member states. EFSA Support Publ 8(3):127E

EFSA (2014) Guidance on the EU Menu methodology. EFSA J 12(12):3944

EFSA (2019) The raw primary commodity (RPC) model: strengthening EFSA's capacity to assess dietary exposure at different levels of the food chain, from raw primary commodities to foods as consumed. EFSA Support Publ 16(1):1532E

EFSA, Dujardin B, Bocca V (2019) Cumulative dietary exposure assessment of pesticides that have chronic effects on the thyroid using SAS® software. EFSA J 17(9):e05763

Eldridge A, Piernas C, Illner A-K, Gibney M, Gurinović M, de Vries J, Cade J (2019) Evaluation of new technology-based tools for dietary intake assessment—an ILSI Europe dietary intake and exposure task force evaluation. Nutrients 11(1):55

Finocchiaro G (2010) Anonimato e Identità Personale (diritto alla), in AA.VV. Digesto delle discipline privatistiche, Sez. Civ. Agg. Utet, Torino, 721

Fischler C (1988) Food, self and identity. Inf (Int Soc Sci Council) 27(2):275–292

Floridi L (1999) Information ethics: on the philosophical foundation of computer ethics. Ethics Inf Technol 1(1):33–52

Floridi L (2005) The ontological interpretation of informational privacy. Ethics Inf Technol 7(4):185–200

Floridi L (2006) Four challenges for a theory of informational privacy. Ethics Inf Technol 8(3):109–119

Floridi L (2013) The ethics of information. Oxford University Press, Oxford

Floridi L (2017) Group privacy: a defence and an interpretation. In: Group privacy. Springer, Berlin, pp 83–100

Fried D, Surdeanu M, Kobourov S, Hingle M, Bell D (2014) Analyzing the language of food on social media. In: 2014 IEEE international conference on big data (big data). IEEE, Piscataway, pp 778–783

FTC, Federal Trade Commission (2015). Is aggregate data always private? https://www.ftc.gov/news-events/blogs/techftc/2012/05/aggregate-data-always-private. Accessed 22 April 2022

Giabbanelli P, Adams J (2016) Identifying small groups of foods that can predict achievement of key dietary recommendations: data mining of the UK national diet and nutrition survey, 2008–12. Publ health Nutrit 19(9):1543–1551

Gilsenan MB (2015) Data handling: observatories-databases-data storage-legal framework EFSA data collection. In: Options Méditerranéennes. Series A: mediterranean seminars. CIHEAMI-AMZ, Zaragoza (Spain)-EFSA, European Food Safety Authority, Parma

Hearty AP, Gibney MJ (2013) Dietary patterns in Irish adolescents: a comparison of cluster and principal component analyses. Publ Health Nutrit 16(5):848–857

Hu FB (2002) Dietary pattern analysis: a new direction in nutritional epidemiology. Current Opin Lipidol 13(1):3–9

ICO (2017) Big data, artificial intelligence, machine learning and data protection. https://ico.org.uk/media/for-organisations/documents/2013559/big-data-ai-ml-and-data-protection.pdf

Lazarou C, Karaolis M, Matalas A-L, Panagiotakos DB (2012) Dietary patterns analysis using data mining method. An application to data from the CYKIDS study. Comput Methods Progr Biomed 108(2):706–714

Lynskey O (2014) Deconstructing data protection: the added-value of a right to data protection in the EU legal order. Int'l Comp LQ 63:569

Malgieri G (2016a) Ownership of customer (big) data in the European union: quasi-property as comparative solution? J Int Law 20(5):37

Malgieri G (2016b) Property and (intellectual) ownership of consumers' information: a new taxonomy for personal data. Privacy Germany-PinG (4):133–150

Malgieri G, Comandé G (2017) Sensitive-by-distance: quasi-health data in the algorithmic era. Inf Commun Technol Law 26(3):229–249

Marcos SV, Rubio MJ, Sanchidrián FR, de Robledo D (2016) Spanish national dietary survey in adults, elderly and pregnant women. EFSA Suppor Publ 13(6):1053E

Marin J, Biswas A, Ofli F, Hynes N, Salvador A, Aytar Y, Weber I, Torralba A (2019) Recipe1M+: a dataset for learning cross-modal embeddings for cooking recipes and food images. IEEE Trans Pattern Anal Mach Intell 43:187–203

Maringer M, van't Veer P, Klepacz N, Verain MCD, Normann A, Ekman S, Timotijevic L, Raats MM, Geelen A (2018) User-documented food consumption data from publicly available apps: an analysis of opportunities and challenges for nutrition research. Nutr J 17(1):1–13

Martoni M, Palmirani M (2015) Internet e identità personale. Filosofia del diritto e nuove tecnologie. Prospettive di ricerca tra teoria e pratica, pp 295–308

Martoni M, Pattaro E (2021) Le idee di identità e di identità personale nei presocratici e in diritto italiano. Giappichelli Editore

Moor JH (1991) The ethics of privacy protection. Library Trends, 39(1 & 2):69–82

Moor JH (1997) Towards a theory of privacy in the information age. ACM Sigcas Comput Soc 27(3):27–32

Nissenbaum H (2020) Privacy in context. Stanford University Press, Redwood City

Ocké M, de Boer E, Brants H, van der Laan J, Niekerk M, van Rossum C, Temme L, et al (2012) PANCAKE – pilot study for the assessment of nutrient intake and food consumption among kids in Europe. EFSA Support Publ 9(9):339E. https://doi.org/10.2903/sp.efsa.2012.EN-339. eprint: https://efsa.onlinelibrary.wiley.com/doi/pdf/10.2903/sp.efsa.2012.EN-339

Ohm P (2009) Broken promises of privacy: responding to the surprising failure of anonymization. UCLA l Rev 57:1701

Palmirani M, Martoni M (2006) Il cittadino elettronico e l'identità digitale nell'egovernance. Gedit

PDPC, Personal Data Protection Commission Singapore (2018) Guide to basic data anonymisation technique. https://www.pdpc.gov.sg/-/media/Files/PDPC/PDF-Files/Other-Guides/Guide-to-Anonymisation_v1-(250118).pdf. Accessed 22 April 2022

Pino G (2010) L'identità personale. S. RODOTÁ: P. ZATTI (eds.). Trattato di biodiritto 1:297–321

Public Health, National Institute for the Environment, van Rossum C, Nelis K, Wilson C, Ocké M (2018) National dietary survey in 2012–2016 on the general population aged 1–79 years in theNetherlands. EFSA Suppor Publ 15(9):1488E

Purtova N (2018) The law of everything. Broad concept of personal data and future of EU data protection law. Law Innov Technol 10(1):40–81

Rigdon J, Basu S (2019) Machine learning with sparse nutrition data to improve cardiovascular mortality risk prediction in the USA using nationally randomly sampled data. BMJ Open 9(11):e032703

Rosso N, Giabbanelli P (2018) Accurately inferring compliance to five major food guidelines through simplified surveys: applying data mining to the UK national diet and nutrition survey. JMIR Public Health Surveillance 4(2):e56

Rozin P, Hammer L, Oster H, Horowitz T, Marmora V (1986) The child's conception of food: differentiation of categories of rejected substances in the 16 months to 5 year age range. Appetite 7(2):141–151

Sasahara K (2019) You are what you eat. J Comput Soc Sci 2(2):103–117

Schwartz PM, Solove DJ (2011) The PII problem: privacy and a new concept of personally identifiable information. NYUL Rev 86:1814

Sette S, Le Donne C , Piccinelli R, Arcella D, Turrini A, Leclercq C, INRAN-SCAI 2005–06 Study Group et al (2011) The third Italian national food consumption survey, INRAN-SCAI 2005–06-part 1: nutrient intakes in Italy. Nutr Metab Cardiovasc Dis 21(12):922–932

Stepanov I (2020) Introducing a property right over data in the EU: the data producer's right-an evaluation. Int R Law Comp Technol 34(1):65–86

Tamò-Larrieux A (2018) Mapping the privacy rationales. In: Designing for privacy and its legal framework. Springer, Berlin, pp 27–43

Tavani HT (2008) Floridi's ontological theory of informational privacy: some implications and challenges. Ethics Inf Technol 10(2–3):155–166

Tavani HT, Moor JH (2001) Privacy protection, control of information, and privacy-enhancing technologies. ACM Sigcas Comput Soc 31(1):6–11

Taylor L, Floridi L, Van der Sloot B (2016a) Group privacy: new challenges of data technologies, vol 126. Springer, Berlin

Taylor L, Floridi L, Van der Sloot B (2016b) Introduction: a new perspective on privacy. In: Group privacy. Springer, Berlin, pp 1–12

Tene O, Polonetsky J (2012) Big data for all: Privacy and user control in the age of analytics. Nw J Tech Intell Prop 11:xxvii

Thompson FE, Subar AF, Loria CM, Reedy JL, Baranowski T (2010) Need for technological innovation in dietary assessment. J Acad Nutr Dietet 110(1):48–51

Upton G, Cook I (2014) A dictionary of statistics 3e. Oxford University Press, Oxford

Van den Puttelaar J, Verain MCD, Onwezen MC (2016) The potential of enriching food consumption data by use of consumer generated data: a case from RICHFIELDS. In: Proceedings of measuring behavior 2016

Yakovlev PA, Guessford WP (2013) Alcohol consumption and political ideology: What's party got to do with it? J Wine Econ 8(3):335–354

Zencovich VZ (1993) Identità personale. UTET, Torino

Chapter 4
Current Trends, Machine Learning, and Food Safety Data Governance

4.1 Data and Algorithmic Transparency

Algorithms are placed outside the legal discourse regarding food safety risk assessment. The scope of transparency and openness as regards the use of computer programmes that allow advanced analytics techniques is unknown, as neither related provisions in the amended legal framework nor case law from European Courts can be safely applied. The relationship between computer programmes and environmental information following Courts' interpretation of transparency has already been introduced in Sect. 2.3.1, but several issues are left without answer. Focusing on algorithms and broadening the discussion to include emerging trends in law and technology, it is necessary to understand the implications of the novel manners in which we 'make sense' of data.[1]

Foreseeable legal challenges in this domain necessarily revolve around the growing use of machine learning algorithms in the context of risk assessment. In particular, future challenges for transparency—yet, unsolved by the Transparency Regulation—regard the degree of institutional access to algorithms and statistical models used by the industry when drafting dossiers and, at the same time, the general, reactive or proactive access to the models used by EFSA in its risk assessment, for instance in automated literature reviews.

On the one hand, an area of investigation should aim to address the extent to which, in light of the principles of openness and transparency, algorithms and machine learning models should be made available to the public in a way that is necessary and proportionate to the consequential limitation of intellectual property rights.[2] On the other hand, a natural progression of 'algorithmic transparency'

---

[1] See the foundational work by Barocas and Selbst (2016)

[2] De Minico (2019) observes that '[a]s regards the functional transparency of the algorithm, it is satisfied by the presence of a selective disclosure, that covers only the most relevant features of the algorithm, to allow interested parties to understand the goals of the predictive mechanism. A

S. Sapienza, *Big Data, Algorithms and Food Safety*, Law, Governance and Technology Series 52, https://doi.org/10.1007/978-3-031-09367-8_4

studies should see how to govern their use in the domain at stake, especially when probabilistic models are used to draft dossiers submitted to EFSA in support of applications for regulated products. Finally, proposed regulations on data governance call for a discussion towards the future of data in the domain of this book. This set of seemingly heterogeneous issues presents the common trait of being rising foreseeable concerns that relate to data governance in food safety and the way they are analysed.

Section 4.1 describes the technicalities underlying the paradigm shift from deterministic to probabilistic (or stochastic) approaches, then contextualises the emerging problem of biased analysis and introduces the concept of 'Knowability' as a viable avenue to mitigate such issues.

Section 4.2 discusses the implications of possible fallacies from an ethical and legal point of view by stressing the relevance of fairness in risk assessment and the relationship between machine learning and the precautionary principle.

Section 4.3 presents some preliminary reflections on the impact of the proposed EU Data Governance Act on food safety data governance.

4.1.1 The Shift from Deterministic to Stochastic Analysis

Alongside traditional (deterministic) methodologies, new paradigms involving the use of probabilistic modelling are emerging. Areas such as the automatic review of the literature, exposure assessment, and outbreaks monitoring, are some of the most affected domains. This subsection, rather than providing an in-depth literature review, discusses some prominent examples of novel trends.

EFSA uses a wide range of tools to perform risk assessment. For instance, the Pesticide Residue Intake Model (PRIMo) model for pesticide chronic and acute exposure assessment[3] is based on equations grounded on scientific evidence. The model uses food consumption data, while calculation is performed by an Excel file freely available on EFSA website.[4] In this evaluation, no machine learning technique is used.

Alongside simple spreadsheets like PRIMo, EFSA has also commissioned studies to evaluate probabilistic approaches to risk assessment. While a relevant part of its research already aims to forecast future trends, the novelty of this approach consists of the adoption of probabilistic analytical methods. It is crucial to point out the differences between traditional modelling and emerging Machine

full knowledge of the algorithms would unduly erase intellectual property rights granted to its legitimate owner, without contributing to the goal of the disclosure, as making available to the public the whole functioning of the algorithm would only benefit expert people. Therefore, this extreme position would hamper without creating any advantage: it has to be rejected due to the lack of necessity and proportionality' (*ed.* author's own translation. Original language: Italian)

[3] EFSA (2018)

[4] See https://www.efsa.europa.eu/en/applications/pesticides/tools, accessed on 27/04/2022

Learning'approaches. In both cases, the output of the calculation consists of a likelihood. While traditional models, given the same initial conditions, generate the same output, machine learning models do not share this property and embed some elements of stochasticity in the final model.[5] Following Russell and Norwig's definitions,[6] we will refer to the traditional statistical modelling and rule-based expert system as 'deterministic', whereas the machine learning approaches will be broadly identified as 'stochastic' or 'probabilistic'.

For instance, a recently developed methodology regarding chemical contaminants exposure assessment relies on Monte Carlo simulation.[7] This methodology makes use of random numbers to perform a given set of simulations in which exposures are calculated on the basis of observations of contaminated foods and food consumption data. The same attempt has been made with cumulative dietary exposure assessment of pesticides[8] and, when compared to the 'SAS' software,[9] encouraging results on the reliability of probabilistic models have emerged: authors observed that 'Comparison of the results revealed that both tools produced nearly identical results and any observed differences are mainly attributed to the random effect of probabilistic modelling'.[10]

EFSA's 'probabilistic turn' is even more evident in the commissioned report 'Machine Learning Techniques applied in risk assessment related to food safety',[11] which proposes both an extensive literature review and a decision tree for adopting machine learning techniques *vìs-a-vìs* deterministic models in the course of risk assessment. Following the literature review and the selection of machine learning techniques that could be deployed by EFSA, two different analyses have been performed, namely (a) a comparison between deterministic and probabilistic approaches, and (b) five case-studies in which stochastic methods have been deployed.

EFSA's attempts towards the adoption of machine learning techniques are not isolated. For instance, the use of probabilistic models for exposure assessment was theorised already in early 2000s', but the absence of well-defined collection methodologies was seen as an obstacle[12,13]. Today, the availability of Big Data and sufficient computational power makes possible the adoption of these techniques also

[5] This is a vital feature of machine learning models that characterise their advantages over traditional approaches. Stochasticity allows a high degree of accuracy towards new data by refining the model over time when new training data are added

[6] Russell and Norvig (2010), p. 43

[7] AGES et al. (2019)

[8] Klaveren et al. (2019)

[9] EFSA et al. (2019)

[10] EFSA et al. (2019), p. 32

[11] IZSTO et al. (2017)

[12] See Lambe (2002)

[13] This article is significant for its historical relevance, rather than its effective contribution to our discussion. Other issues, such as the availability of data and the computational resources needed for the deployment of machine learning techniques were not fully addressed

by academic research which becomes part of the literature used by EFSA for risk assessment purposes.

A study[14] has proposed a comparison between logistic regression and random forest to analyse causal relationships between the exposure to certain risks (i.e. direct contact with animals, consumption of raw/uncooked meat) and illnesses, even in case of missing values or high number of possible exposures. Random forest algorithms were proved to be more effective than logistic regression. Similarly, a short research[15] has shown the efficacy of combining of on-farm (animal movement) and environmental variables to generate a classification tree (C4.5 algorithm) capable of describing factors for risk profiles in cattle herds.

Pesticide residue monitoring can benefit from machine learning.[16] The combination of electric biosensors and Artificial Neural Networks (ANNs) has reached a sufficient degree of accuracy in screening different kinds of residue in pesticide monitoring.[17] If scalable, this system might increase the number of tests carried out in monitoring programmes.

An American study on environmental factors impacting vegetables and fruit[18] is particularly significant due to the variety of data and machine learning techniques deployed. Authors included data on observed samples, topographical and spatial data to automatically generate a classification tree helpful to predict the presence/absence of food-borne pathogens linked to mereological or topographical conditions. Image recognition can be combined with digital microscopy or a faster and less expensive detection of parasites. For instance, this has been the case of *Eimeria* detection in domestic chicks using Bayesian classifier.[19] Support Vector Machine and Artificial Neural Networks have been used to recognise and classify species of harmful beetles.[20]

In summary, the 'probabilistic turn' of food safety comprises several techniques approached by EFSA and consolidated academic research. Let us analyse their broader impact in the next section.

---

[14] Gu et al. (2015)

[15] Ortiz-Pelaez and Pfeiffer (2008)

[16] In compliance with Regulation (EC) No 396/2005 of the European Parliament and of the Council of 23 February 2005 on maximum residue levels of pesticides in or on food and feed of plant and animal origin and amending Council Directive 91/414/EEC [2005] OJ L 70/1 and implementing regulations (e.g. Commission Implementing Regulation (EU) 2017/660) Member States are obliged to develop country-wide pesticide monitoring programmes to ensure that rules concerning maximum residue levels are respected

[17] Ferentinos et al. (2013)

[18] Strawn et al. (2013)

[19] Castañón et al. (2007)

[20] Bisgin et al. (2018)

4.1.2 *Possible Biases in Food Consumption Data Analysis*

Once clarified the shift towards probabilistic approaches, it is possible to discuss some of their implications. For the purposes of this study, we will refer to 'Data fallacies' as mistakes in data collection, analysis or interpretation. They represent a major challenge for every domain affected by the ongoing datafication process.[21] These kinds of errors can be classified according to three categories:

- Data-driven fallacies. For instance, a wrongful construction of training sets for machine learning algorithms might originate sampling biases, incompleteness due to selection criteria ('survivorship bias'), or manipulation of data grouping criteria ('gerrymandering').
- Analysis-driven fallacies. Underfitting and overfitting of statistical models is a well-known issue of machine learning modelling that pertains to the under- or over-representativeness of the statical model in relation to the training data. When underfitting, the model lacks of a sufficient descriptive power to identify a generalisable pattern. Overfitting, instead, consists of the poor generalisability of the model due to its extreme adherence to the training data. 'Regression towards the mean' mistake occurs when phenomena are generalised using statistical mean in order to prevent peaks or unexpected results.
- Interpretation-driven fallacies. 'False causality' is a wrong assumption of cause-effect relationship for trends that simply present spurious similarities; relying on summary metrics might hamper the significance of differences in raw data; 'cherry picking' (or confirmation bias) is the practice of selecting results that fit certain claims while excluding others; finally, the so-called 'Gambler's fallacy' is the trend to believe that something unusual according to the data will or will not happen in the future

Let us discuss some significant examples by means of a theoretical experimental framework.[22] In the following Fig. 4.1, five stages are displayed. First some data constitute the training set of one (ore more) supervised machine learning algorithm(s) used to obtain results. Its outputs are transposed in a scientific paper similar to the ones mentioned in our technical review above, thus framed as a

---

[21] The literature has also discussed data fallacies in terms of 'bias'. The literature on this issue is abundant. *Inter alia*, see the foundational work by Mittelstadt and Floridi (2016) and a recent reply Tsamados et al. (2022). See also Bellamy et al. (2019) Several legal scholars have discussed the issues of algorithmic bias. See, *ex multis*, perspectives on constitutional law Micklitz et al. (2021), on criminal law Barabas (2020), consumer law Hacker (2018), and legal philosophy Palmirani and Sapienza (2021). See also further references in Sects. 4.1.1 and 5.2.3 on the notion and the principle of Fairness

[22] Unfortunately, the literature on emerging biases in food safety information retrieval systems does not seem abundant. A close domain might be the one of social biases on information retrieval, discussed in short papers (e.g., Otterbacher 2018, or in word embedding Papakyriakopoulos et al. 2020). Therefore, I will proceed with a thought experiment trying to integrate the literature in algorithmic bias with the case studies performed at EFSA

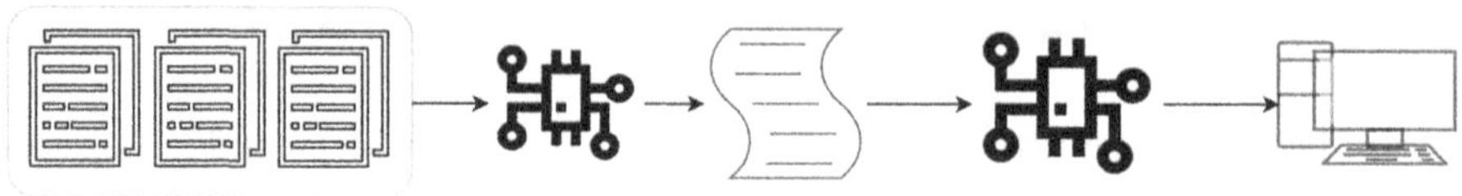

Fig. 4.1 Automation of literature review

classification problem (relevant / non-relevant). Then a final user—let us say, a risk assessor—retrieves the paper by means of a machine learning algorithm that performs an automated literature review.[23] Scientific papers, in the last stages, constitute training sets for the algorithms which perform the literature review.

The reality is, of course, far more complex than scenario displayed above. In fact, the data-chain described so far might include several agents that make use of machine learning techniques. This hypothetical setting is likely to raise concerns regarding the propagation of inherent model biases that could be spread to other studies and reach the 'evidence' level in EFSA opinion. The pictures below show some noteworthy cases of biases.

Figure 4.2 shows three non-biased data-chains (A, B, D). In these chains, colours in the training sets represent some degree of diverse input data (e.g. food consumption data belonging to diverse ethnic groups). Instead, chain C does not presents the same amount of diversity *within* itself. This situation might emerge as a consequence of the aforementioned data-driven fallacies. Line C algorithm generates results that are transposed into a biased paper: despite being correct from a mathematical perspective, its results might be fallacious if we take into account that certain food patterns are under-represented. Paper C propagates the bias to the final user by being embedded in the model generated by the machine learning algorithm that performs the automatic reviews. This situation will be referred to as Type A bias.

This scenario could also occur in the absence of machine learning models deployed across the data-chain. However, their presence considerably increases the risks associated to Type A bias as some of the models used in the scenario might not allow an in-depth scrutiny of their functioning (the so-called 'Black Box' problem[24]) or if training data are not made available to the final user and they can only trust the overall results (e.g. precision and recall).

One might think that the problem of Type A bias propagation could be solved by adding some degree of diversity *within* paper C. Figure 4.3 shows that, in reality, another kind of bias might emerge, for instance if selected papers come from the same author. This is the case of absence of diversity *among* rather than *within* papers that are automatically scrutinised. This scenario will be named after Type

[23] IZSTO et al. (2017)

[24] Pasquale (2015)

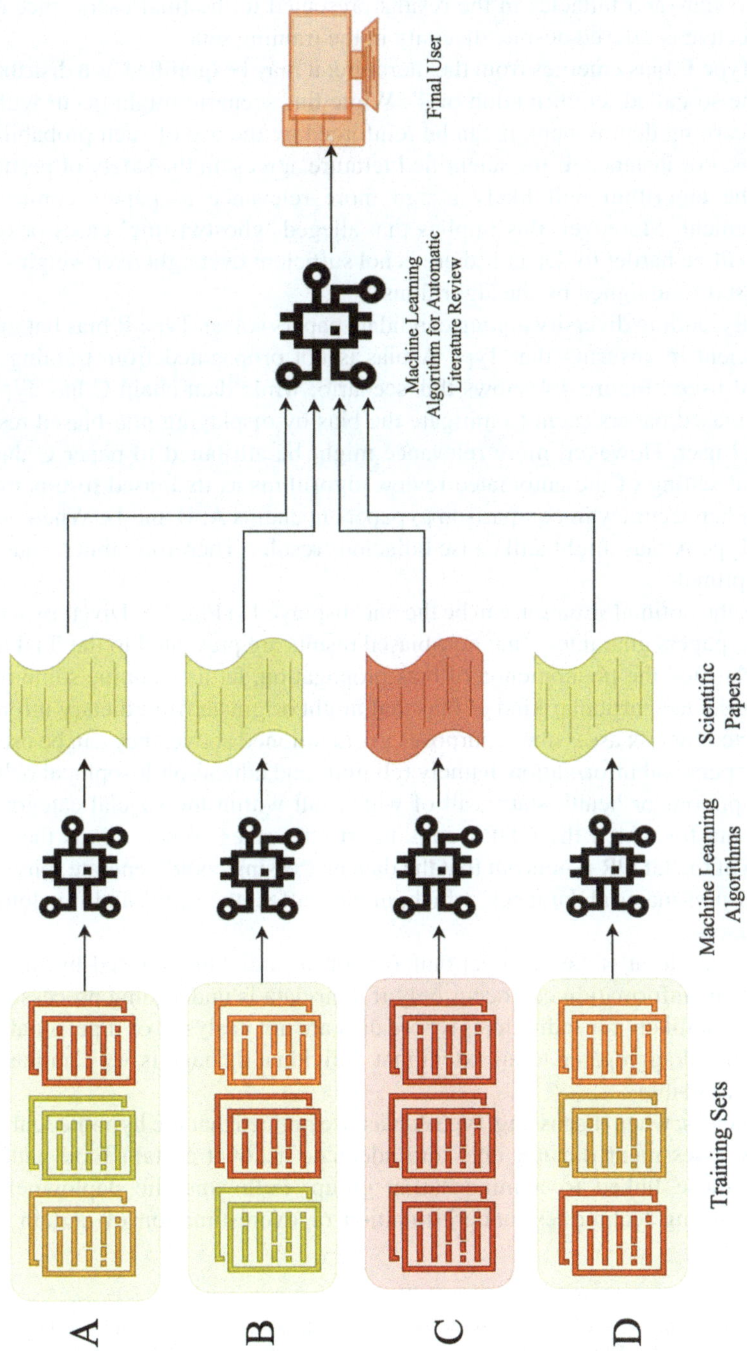

Fig. 4.2 Bias propagation—Type A bias

B bias. Its consequences are the lack of diversity in candidate papers selected for literature review and fallacies in the results presented to the final users since only one perspective is offered despite diversity in the training sets.

Since Type B bias emerges from the literature, it may be qualified as a distributed form of the so-called 'confirmation bias'. While this scenario might occur without machine learning deployment, it can be reinforced by the use of such probabilistic approaches. For instance, if the scientific literature agrees on the safety of pesticide *X*, then the algorithm will likely assign more relevance to papers confirming such agreement. Moreover, this implies that alleged 'ghostwriting' cases or other scandals will be harder to detect if there is not sufficient oversight over weights and relevance scores assigned by the algorithms.

Crucially, adding diversity *among* candidate papers solves Type B bias but might be insufficient in ensuring that Type A bias is not propagated from training sets to the final users. Figure 4.4 shows this scenario: while data-chain C has Type A bias, non-biased papers seem to mitigate the bias by displaying non-biased results to the final user. However, more relevance might be attributed to paper C due to the internal setting of the automated review algorithms as its biased results might display higher accuracy in comparison to papers in chains A, B and D. When this is the case, Type A bias might still cause fallacious results. Therefore, this scenario is still sub-optimal.

Finally, the optimal situation can be the one displayed in Fig. 4.5. Diversity *within* and *among* papers guarantees that non-biased results are presented to the final user.

Once clarified the phenomenon of bias propagation, let us examine some of its implications. One particular kind of bias that might originate from dietary information collected for risk assessment purposes. As mentioned above, they can be used to infer other personal information, namely religious and ethical/philosophical beliefs, political opinions or health status, all of which fall within the special category of personal data for which the GDPR calls for stricter rules. When this is the case, Recital 75 of the GDPR points out that the data processing could generate 'physical, material or non-material damages' which, in turn, affect the rights and freedoms of natural persons.

In the scenario at stake, a direct damage for an individual caused by the processing of this information can occur only if their data is undergoing processing,[25] whereas it is unlikely for individuals whose data are not analysed or if personal data are anonymised/aggregated to an extent that individual damage is very unlikely or virtually impossible.

Nevertheless, when discussing Type A bias, we noticed that the hypothetical food safety risk assessment did not take into adequate account certain food patterns, especially those linked to a minoritarian group. Following the deployment of machine learning techniques, underestimation or overestimation of certain food

[25] For instance, in the event of a data breach (Recital 85). To assess the magnitude of the risks, it is worth mentioning that EFSA food consumption database contains $\simeq$ 100.000 individual records that aim to generalise the behaviour of more than 500 million individuals

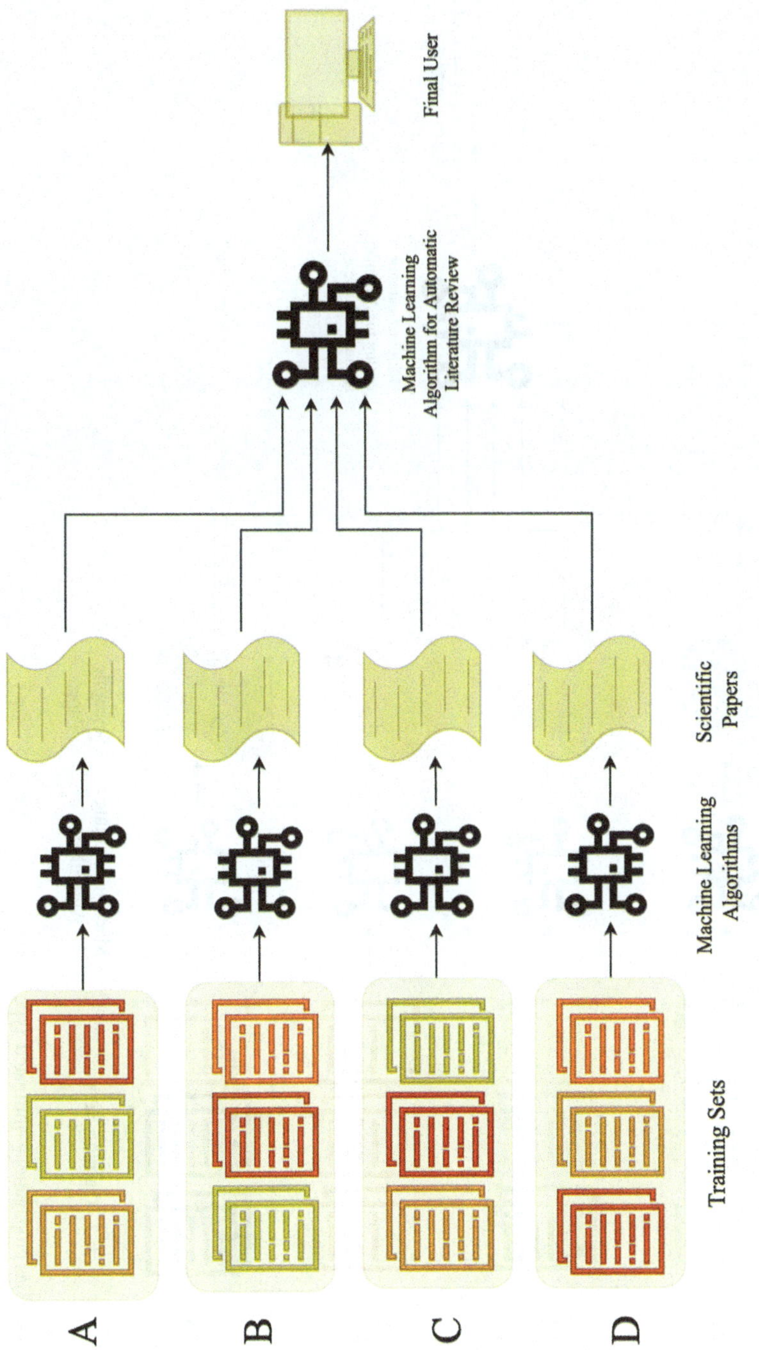

Fig. 4.3 Bias propagation—Type B bias

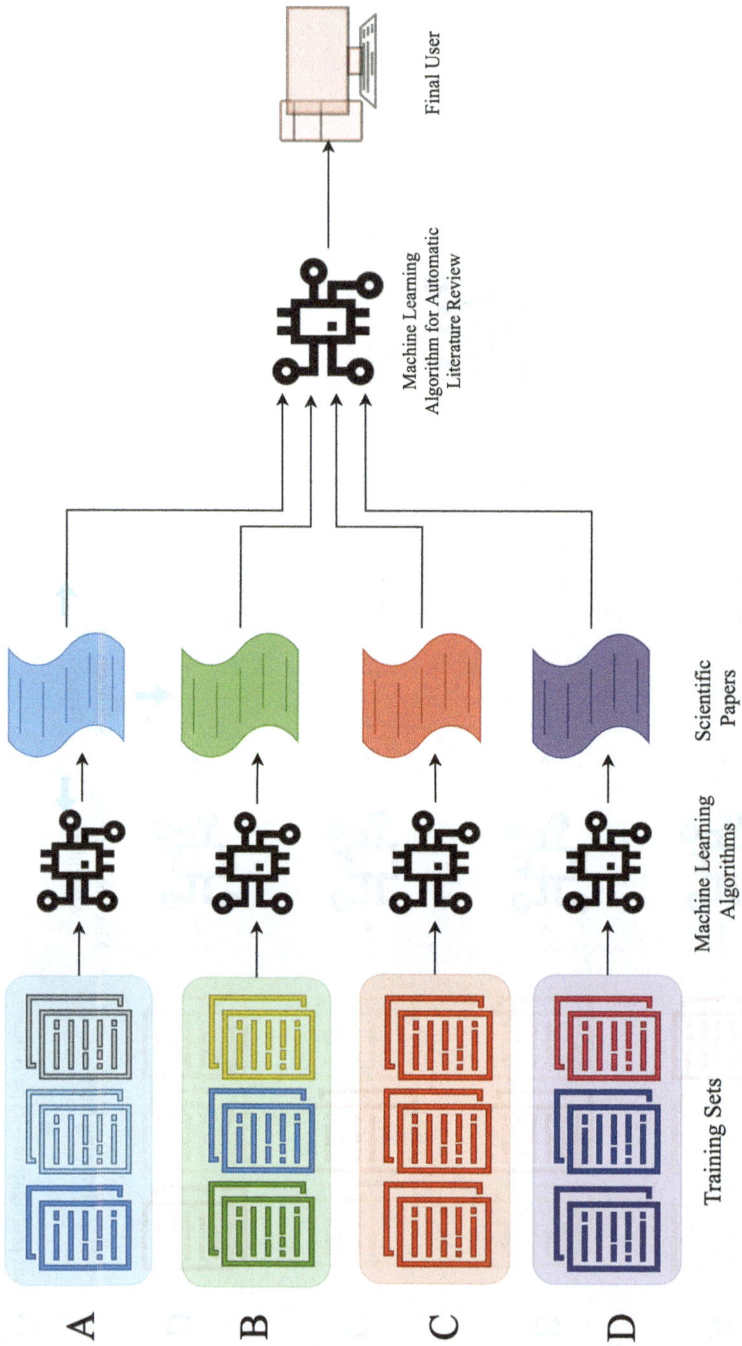

Fig. 4.4 Bias propagation—mitigated Type A bias

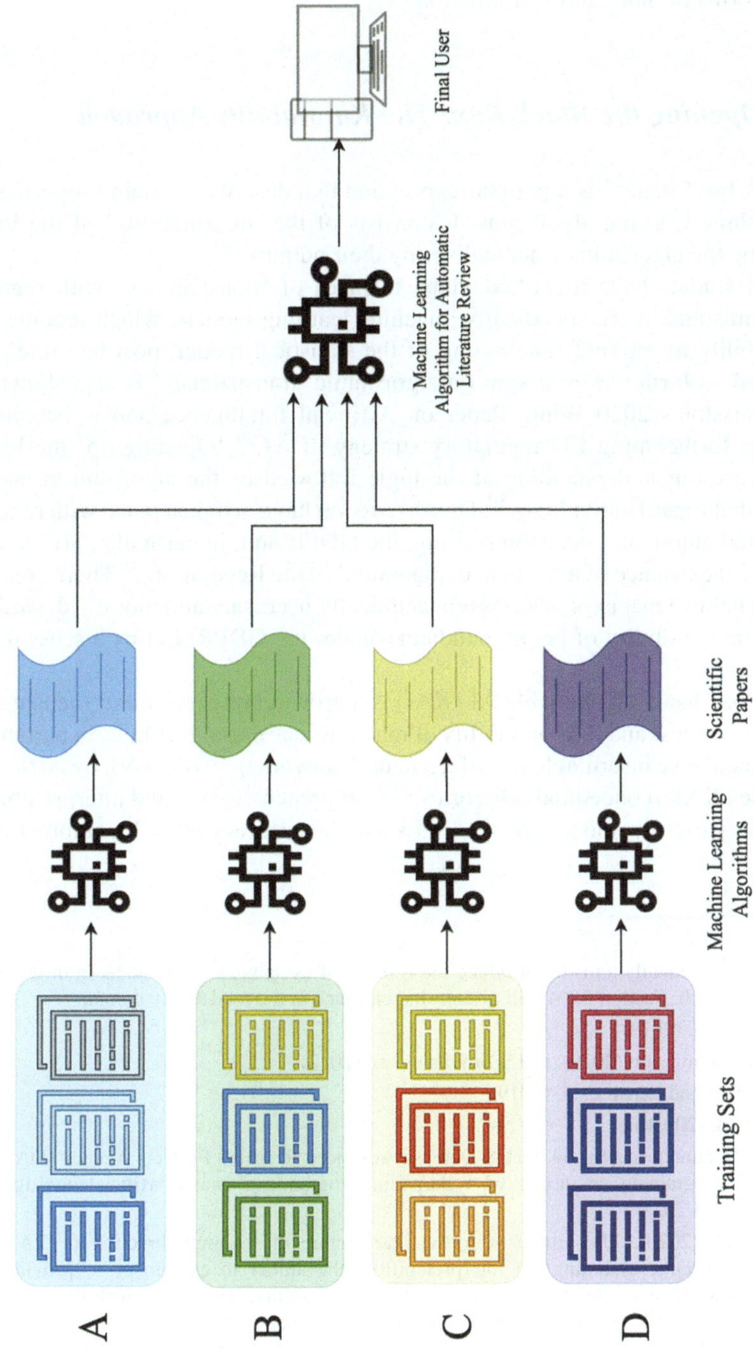

Fig. 4.5 Bias propagation—absence of bias

patterns in input datasets (in particular, in supervised approaches) may thus lead to subtle forms of indirect discrimination.

4.1.3 Opening the Black Box: The Knowability Approach

The 'black box' issue[26] is a popular expression that describes certain properties of some machine learning algorithms. It consists of the 'inscrutability' of the logic followed by the algorithms when delivering their outputs.[27]

Several studies have identified many nuances of 'transparency' with regards to algorithms and, more specifically, machine learning models, which include the interpretability of internal functioning of the statistical model, post-hoc interpretations, and so forth.[28] Discussion on algorithmic 'transparency' is also shown in the Commission's 2020 White Paper on Artificial Intelligence and is becoming part of the forthcoming EU regulatory strategy of AI.[29] 'Opening up' the black box will ease our understanding of the logic followed by the algorithm in taking decisions that regard individuals.[30] This discussion has also taken place with regards to individual automated-decision making, the GDPR and, in particular, art. 22 and the contested existence of a 'right to explanation' in the Regulation.[31] There are also reasons to believe that explanations are technically necessary and morally desirable, let alone the possibility of being mandatory under the GDPR. Let us discuss these perspectives.

On the one hand, eXplainable AI (XAI) is a growing research trend focusing on comprehensiveness and interpretability of machine learning algorithms, in particular those deemed to be inscrutable (e.g. deep neural networks).[32] The work by Arrieta et al. identifies XAI's conceptual categories.[33] Comprehensiveness and interpretability are necessary to truly understand the limits and the accuracy of such computational

---

[26] Pasquale (2015)

[27] This subsection mainly aims to introduce the concept of explainable AI systems in informatics and ethical research. Section 5.2.4 will discuss its relevance in the food safety domain

[28] Mittelstadt et al. (2019)

[29] European Commission (2020), p. 15, Sovrano et al. (2022)

[30] Zarsky (2016) and Zerilli et al. (2019)

[31] Wachter et al. (2017b)

[32] See the 2016 announcement by the US Defense Advanced Research Projects Agency (DARPA) on its XAI programme at https://www.darpa.mil/program/explainable-artificial-intelligence, accessed on 27/04/2022

[33] Arrieta et al. (2020). The article offers a clear picture of research direction in XAI and fundamental concepts, including (a) Interpretability, 'the ability to explain or to provide the meaning in understandable terms to a human'; (b) Explainability, 'associated with the notion of explanation as an interface between humans and a decision maker that is, at the same time, both an accurate proxy of the decision maker and comprehensible to humans', and (c) Transparency, 'according to which a model is considered to be transparent if by itself it is understandable'

approach. For instance, Local Interpretable Model-Agnostic Explanations (LIME) is an attempt to understand the prediction of image classifiers.[34] Various proposals regarding the properties and the content of explanations, both for the purposes of the GDPR and other contexts, have been made.[35]

On the other hand, in parallel with XAI research trend, authors in the field of AI and data ethics have argued in favour of the moral desirability of explanations. The AI4People working group has proposed the 'Explicability' principle by correlating the intelligibility of AI systems to the 'accountability' for their use.[36] This dimension is crucial to understand the social relevance of AI outputs. By ensuring that explanations are timely provided to the users on the basis of their understanding of reality, AI systems can become trustworthy.[37] More recently, the discussion has also included the proposed AI Act.[38]

While transparency is a well-known principle governing data transmission among the stakeholders active in the food safety systems, its contextualisation regarding algorithms necessarily implies some kind of convergence between the general concepts of algorithmic transparency and the meaning attributed within food safety. This operation requires an in-depth scrutiny of similarities and differences between these concepts.

First, food safety transparency mainly revolves around data, with few exceptions.[39] While broadening the scope of 'environmental information' has been proposed before to include algorithms and operational data (Sect. 2.3), the regulatory framework seems reluctant in accepting such broad notions. Instead, XAI and explicability are mostly algorithmic-centric because they directly tackle issues that relate to computations and the final outputs.

Second, obstacles to transparency have different nature. Black box issues are understood as the lack of intelligibility of the algorithm, whereas obstacles to transparency in food law are mainly due to commercial interests of private stakeholders. While algorithmic transparency may also be hindered by intellectual property laws or under other legal obligations (e.g. contracts), explainability as a research trend is mainly understood as a technical challenge rather than a legal issue.

Third, both notions of transparency are *functional* to achieve a given goal. Food law transparency is multi-faceted, but in our previous reconstruction it has been deemed functional to increase trust. The same goes with explanations of machine learning algorithms when considered as a principle (explicability) in the light of

---

[34] Ribeiro et al. (2016a) and Ribeiro et al. (2016b)

[35] Sovrano et al. (2019) and Miller (2019)

[36] AI4People (2018)

[37] High-Level Expert Group on Artificial Intelligence (2019)

[38] See, also in relation to the possibility of identifying metrics to measure the level of explainability Sovrano et al. (2022)

[39] The submission of 'the programming code used for the statistical analysis' for the comparison between the genetically modified plant and its conventional counterpart is mandatory (point 1.3.2.2 of Annex II of the Commission Implementing Regulation 503/2013 implementing Regulation 1829/2003 on Genetically Modified food and feed)

ensuring trustworthiness of AI systems according to institutional documents at EU level.

Therefore, while differences between the kind of transparency in food law and algorithmic transparency, trust can serve as a unifying perspective to discuss for a joint discussion. However, as differences between definitions persist on the objective scope (data vs algorithms) and nature of the issue to be tackled (technical vs legal) none of the approaches is entirely satisfactory and a broader reflection seems to be needed.

An attempt has been made by scholars discussing algorithmic transparency in the context of public decision-making. This approach focuses on the social acceptability of the decisions taken by automated means by highlighting the role of the perception by the public. From these premises, transparency affects the public's perception of the decision in a manner to provide sufficient legitimacy. When this is the case, these authors suggest that transparency shall focus on the *rationale* underlying the decision rather than the *process* through which the decision has been taken, consistently with the principle of explicability.[40] While appropriate for this context, their approach proceeds to clarify *how* explanations can work in the public decision-making processes. Instead, our reflection mainly revolves around the *why* the interpretation of transparency in food law and explicability can co-exist. To find this answer, the perceived legitimacy of the decision is significant but not decisive: in food safety system, the decision by the risk manager follows the risk assessment performed by EFSA, that is also the first receiver of the explanation by the algorithm. Nonetheless, the decision by the Commission is based on EFSA's opinions, hence the necessity of discussing its social impact.

Other authors have proposed the concept of 'Knowability'.[41] A joint reading of these two studies suggests that explanations in the GDPR suffer from limits related to (a) the material scope of the GDPR, that does not cover non-personal data, discrimination, and liabilities[42] and (b) that the discussion on explainability shall be expanded to include components other than algorithms, namely data.[43]

Discussing in detail the second knowability approach, it proposes the release of *ex-ante* and *ex-post* elements that constitute the content of an explanation. They include, *inter alia*: mathematical models used to build the algorithm; training, development, and testing datasets for machine learning models; metadata and contextual information about the data used; ontologies used to represent the knowledge in the system.[44] As such, this approach goes *beyond* mere algorithmic transparency and encompasses informational components.[45]

[40] de Fine Licht and de Fine Licht (2020), p.919

[41] See Pagallo (2020), Palmirani (2020), and Palmirani and Sapienza (2021)

[42] Pagallo (2020)

[43] Palmirani (2020) and Palmirani and Sapienza (2021)

[44] Palmirani (2020)

[45] Palmirani and Sapienza (2021)

The Knowability approach, enriched by the peculiarities emerging from the food safety domain, can be a unified perspective to discuss food law transparency and explicability holistically. Trust in the food safety system represents a crucial element in the literature on transparency and explicability because, as argued above, they are both functional to achieve it. Knowability could be grounded on the same rationale.

4.2 Accountability and Redress of Probabilistic Models

The question of bias and data fallacies is also linked to the shift from deterministic to probabilistic methods, which brings along concerns pertaining to the accountability of risk assessors for unfair results. Deterministic algorithms are characterised by an equality relationship such that, given the same input and a functioning software, the output will always be the same. Hence, the correctness of deterministic results can be assessed by analysing whether input data were correctly inserted in the system and the algorithm executed all the planned instructions according to the set of commands given by the programmer.

Vice versa, probabilistic modelling is characterised by results that are expressed in terms of likelihood. As such, they always embed a statistical error (e.g. false positives/false negatives) due to their own nature. Differently from deterministic scenarios, such errors can occur despite the use of high-quality input data and the correct functioning of the software. This factor might leave open the questions regarding how these sorts of mistakes and failures could be identified and who would be held accountable for them.

This section analyses issues related to accountability frameworks in food safety system. First, it discusses the concept of fairness as applied to risk assessment; then, it clarifies the key concepts of responsibility, accountability and liability and how they will be used in this context; finally, it proposes a remedy taking into account the precautionary principle.

4.2.1 Fairness in Risk Assessment

When discussing group privacy and discrimination issues, it was established that the discussion on algorithmic results that unduly differentiate individuals belonging to certain groups shall also keep into account the interpretation of the concept of 'fairness'.

Fairness in the public sector algorithmic decision-making became relevant in 2019 following a ruling from the Italian Consiglio di Stato (Council of State),[46] i.e., the Court responsible of the legality of the public administration in Italy.

[46] See the ruling n. 8472/2019 by the Italian Consiglio di Stato

The case concerned the employment of teachers in public education: appellant teachers were subject to automated decision-making procedures that had the effect of assigning them to towns far from their place of residence. While acknowledging the advantages of using algorithmic decisions as regards costs and efficiency of the public sector, the Court critically noted that these decisions are subject to 'fair administration' principles, including non-discrimination, transparency, etc. (para 8.2 of the ruling). Interestingly, data protection principles were not mentioned in the case, which mainly revolved around the use of algorithms in the public sector.

This perspective is unavoidably broader than privacy-only discussion, but it is intertwined to it by at least two factors.

On the one hand, fairness constitutes one data protection principle in the EU and at the international level (art. 5 of the GDPR, art. 4 of Regulation 2018/1725, art. 5(4)(a) of Convention 108+ of the Council of Europe). On the other hand, broadening the discourse to include foreseeable issues with the deployment of machine learning models—including the ones highlighted in the previous section—entails the necessity of discussing unfair results of machine learning computations that use personal data.

Yet, what is fairness? First, an intuitive difference between *latu sensu* fairness and *strictu sensu* fairness can be drawn. *Latu sensu*, fairness mainly concerns the implications of a given decision and how it shall not discriminate individuals. Then, *strictu sensu* fairness is the bundle of legal requirements that is mandated by data protection law when personal data are processed (art. 5 of the GDPR and the other pieces of legislation mentioned above).

Let us first discuss the *latu sensu* interpretation. According to Council Directive 2000/43/EC,[47] discriminations can be direct or indirect. Article 2 of the Directive states that:

(a) direct discrimination shall be taken to occur where one person is treated less favourably than another is, has been or would be treated in a comparable situation on grounds of racial or ethnic origin;
(b) indirect discrimination shall be taken to occur where an apparently neutral provision, criterion or practice would put persons of a racial or ethnic origin at a particular disadvantage compared with other persons, unless that provision, criterion or practice is objectively justified by a legitimate aim and the means of achieving that aim are appropriate and necessary.

Data protection fairness (hence, our *strictu sensu* interpretation) is mainly focused on the first set of issues. The GDPR places fairness in art. 5 akin to lawfulness and transparency. A linguistic analysis performed by Malgieri[48] highlights the many nuances of the notion of fairness under the Regulation. *Inter alia*, the author remarks that 'Fairness in the GDPR goes beyond mere legality (*summum ius, summa*

---

[47] Council Directive 2000/43/EC of 29 June 2000 implementing the principle of equal treatment between persons irrespective of racial or ethnic origin [2000] OJ-L 180/22

[48] Malgieri (2020)

iniuria): it is aimed to the consideration of the substantial circumstances at stake. This means considering expectations of the individuals, effects on the data subjects, actual interests of the parties. Among possible unfair effects we need to include discrimination, but also any other risks to fundamental rights of individuals (e.g. manipulation)'.

This consideration is also grounded on Recital 71 of the GDPR, which specifies that 'in order to ensure fair and transparent processing in respect of the data subject, taking into account the specific circumstances and context in which the personal data are processed, the controller should use appropriate mathematical or statistical procedures for the profiling, implement technical and organisational measures appropriate to ensure, in particular, that factors which result in inaccuracies in personal data are corrected and the risk of errors is minimised, secure personal data in a manner that takes account of the potential risks involved for the interests and rights of the data subject and that prevents, *inter alia*, discriminatory effects on natural persons on the basis of racial or ethnic origin, political opinion, religion or beliefs, trade union membership, genetic or health status or sexual orientation, or that result in measures having such an effect'. The risk of discrimination is echoed in Recital 75 of the GDPR. The AWP29, in its Opinion on Automated Decision-Making, states that "[P]rofiling may be unfair and create discrimination'.

As regards the issue of non-discrimination, EFSA scientific opinions do not produce legal effects on individuals and mostly affect groups.[49] Instead, for what concerns the expectations of the individuals towards their data, access/distribution rules discussed in Chap. 2 are enriched by contractual agreements, whose contents are unknown to the public. This prevents any further scrutiny on data misuses that can harm specific individuals. Finally, the 'elephant in the room' issue with the *strictu sensu* interpretation of fairness is the inapplicability of data protection law in the context of food safety, as discussed previously.

Then, we could think that going beyond data protection law and opting for a *latu sensu* interpretation can increase our chances to handle unfair outcomes. However, we shall keep in mind that data-driven forms of discrimination are particularly problematic due to the three reasons highlighted by Hacker,[50] i.e. their falling outside of the material scope of anti-discrimination law, the higher chances of indirect discrimination, and the placement of the burden to prove the discriminatory nature of the algorithm placed on discriminated individuals. In our scenario, individual discrimination is unlikely to occur: data aggregation prevents the attribution of consumption patterns to specific individual and EFSA's decision have no direct effects on individuals, let alone the inapplicability of data protection law. If *direct* cannot provide satisfactory results, *indirect* discrimination can be seen as a better option. However, Durante noted that only few elements, namely racial or

---

[49] A similar scenario is discussed in Hallinan and de Hert (2017) with regards to conclusions based on genetic information. Authors note that 'The Regulation [...] was drafted on the presumption that the individual, and individual rights, were the primary target of protection'

[50] Hacker (2018)

ethnic origin can trigger the applicability of anti-discrimination legislation, hence its inadequacy to confront with other forms of discriminations, including the ones that originate on different, yet sensitive, grounds.[51]

Certain elements of *indirect* discrimination are nonetheless relevant despite limitations in its remit for the purposes of Council Directive 2000/43/EC. Indirect discrimination occurs when a seemingly neutral element is used to differentiate among individuals. If we broaden the scope of the provisions, which only covers racial or ethnic origin, in the light of the other grounds provided by Recital 71 of the GDPR (racial or ethnic origin, political opinion, religion or beliefs, trade union membership, genetic or health status or sexual orientation), it is possible to discuss discriminations that can emerge from food safety risk assessment.

Vice versa, fairness in data protection law should be conceptualised as inclusive of indirect forms of discrimination. This does not entail that Article 22 of the GDPR can be applicable to EFSA's opinions: neither do they affect individuals, nor they are based solely on automated means. Nonetheless, Recital 71 suggests that, in order to guarantee fairness, data controllers shall act to minimise risks of discrimination based on certain factors (racial or ethnic origin, health conditions, and so on). This shall happen 'taking into account the specific circumstances and context in which the personal data are processed'. While the risk of indirect discrimination is considerably lower than other scenarios where direct discrimination can occur, it is not absent and controllers might be still required to act.

This conceptual operation is not exempt from some limitations. The GDPR is not meant to pursue anti-discrimination and this interpretation may seem arbitrary and highly discretionary. I believe it is not. First, fairness is acknowledged as one of the cornerstones of data protection law (including at the international level), but its scope goes beyond these pieces of legislation. Different would have been the case with—let us say—data minimisation or purpose limitation principles, which are highly influential in data protection but do not interact with other spheres of the relevant legal framework; second, Malgieri's analysis[52] highlights the many nuances of 'fairness' and their wide scope, all converging towards the notion of vulnerability of the data subject. Such wide interpretation should ease its further expansion towards indirect forms of discrimination of a data subject. Third, despite limitations in data protection law (let alone the possibility of re-identification), Recital 71 of the GDPR might leave gaps for an extensive interpretation of discriminatory effects that include indirect discrimination. This can happen by interpreting 'measures having such an effect' in Recital 71 to be understood as indirect forms of discrimination.

[51] Durante (2019), p. 252

[52] Malgieri (2020)

4.2.2 *Responsibility, Accountability, Liability*

Once clarified the notion of fairness, it is necessary to contextualise this interpretation within the food safety system. It has to be preliminary remarked that unfair decisions come as an epistemic consequence of EFSA's scientific opinions. The epistemic chain is valid also when non-discriminatory, yet inaccurate or simply 'bad' decisions are taken. Unfair decisions taken on the basis of opinions drafted with the support of machine learning algorithms might have moral and legal implications that are worth discussing.

Such consequences represent a serious challenge. It is well studied in theoretical frameworks and applied sectors, also from ethical perspectives.[53] To date, food safety system has not received sufficient attention, perhaps due to the relative novelty of machine learning approaches in this context.

According to this research, accountability gaps represent a major challenge and the focus of a recent debate. On the one hand, the deployment of autonomous machine learning systems might lead to situations where no person (physical or legal) is accountable for the use of the system.[54] This challenge is also studied jointly with transparency, fairness, and interpretability of the machine learning systems and leads to intolerable situations for democratic societies.[55] On the one hand, some advocates that machine learning systems can reach a level of autonomy that might lead to their accountability for their use. When this is the case, they could be held liable, also under criminal law.[56]

The latter position is not truly applicable here. Even though autonomy is a pre-condition of the deployment of machine learning systems in the cases seen in the previous section, their degree of independence from human beings is not sufficiently high to pose risks towards 'accountability gaps'. To put it differently, the level of human oversight is still significantly high in the decision-making process and autonomous decisions—which are the pre-condition of the aforementioned literature—are not technically feasible. Therefore, I would cautiously refrain from endorsing the 'robots can be liable' doctrine in favour of a more human-centric approach capable of attributing the legal consequences of machine-supported decision on physical or legal persons. Nonetheless, some aspects of this debate are relevant in our domain. In particular, the issue of how to allocate and distribute the

---

[53] As regards theoretical aspects, see Contissa et al. (2013) and Lagioia and Contissa (2020); On applied cases, see Pagallo (2017), Veale et al. (2018), and Nalepa and Araszkiewicz (2020). On ethical perspectives, see the work by Floridi et al. (2018)

[54] European Parliament, "Civil Law Rules on Robotics—European Parliament resolution of 16 February 2017 with recommendations to the Commission on Civil Law Rules on Robotics (2015/2103(INL))" (P8_TA-PROV(2017)00 51, European Parliament, 2017), (available at http://www.europarl.europa.eu/sides/getDoc.do?pubRef=-//EP//NONSGML+TA+P8-TA-2017-0051+0+DOC+PDF+V0//EN, accessed on 27/04/2022)

[55] Wachter et al. (2017a) and Pajno et al. (2019)

[56] Hallevy (2010)

moral and legal implications of algorithmic-supported decisions is applicable to our domain due to the existing relation between risk assessors and risk managers.

Such relation is both epistemic and legal. Risk manager's (EU Commission) decisions are informed by the risk assessor's (EFSA) opinions: such transmission of knowledge occurs within an institutionalised framework. In a pre-machine learning algorithm, the scenario can be similar to the one depicted below:

$$\text{Data} \rightarrow \text{Risk assessor} \rightarrow \text{Risk Manager}$$

EFSA's institutional task is to extract knowledge from the data and transmit it to the Commission for its decision. Information can be described as the 'making sense' of data thanks to an interpretation made by an epistemic agent.[57] Evidence is a corroborated statement that proof that information is accurate. Their sum corresponds to the extracted knowledge.

$$\text{Data} \rightarrow \text{Knowledge (Information + Evidence)} \rightarrow \text{Decision}$$

In a post-machine learning scenario, EFSA's evidence is based on the information extracted by the data manually or thanks to machine learning algorithm. In the latter case, which will become more frequent in the near future, the cycle will look as follow:

$$\text{Data} \rightarrow \text{ML algorithm} \rightarrow \text{Risk assessor} \rightarrow \text{Risk Manager}$$

From an epistemic perspective, there is no longer consolidated knowledge: the shift from deterministic to probabilistic algorithms necessarily implies that all the extracted information generates an *approximation* of reality, i.e., a possible, probable, or likely statistical model that represents the reality. While information is extracted by the algorithm, the risk assessor corroborates the information and validates is as evidence for the decision.

$$\text{Data} \rightarrow \text{Information} \rightarrow \text{Evidence} \rightarrow \text{Decision}$$

Once clarified this epistemic chain, let us proceed with the reconstruction of its implications under a moral and legal perspective. I will proceed by following the *agent-patient* moral scheme, also used in ethics of information.[58] First, I will use the term *responsibility* to define the active role played by the agent when causing something the patient, either by acting or not. In other words, the proposed understanding of responsibility is the causal link between the agent and patient. Then, the term *accountability* will be used to identify the allocation of the moral consequence of the action on the entities involved. Finally, the notion of *liability*

[57] Durante (2017)

[58] Floridi (2013)

proposed here is the bundle of legal implications (including fines) related to the actions at stake. While the three legal positions usually co-exist within the same entity, which is responsible, accountable and liable at the same time, some exceptions can 'break' this union. For instance, justifications such as self-defence or state of necessity can make a legal entity non-liable, yet responsible from a causal perspective.[59] The separation of responsibility, accountability and liability is also a way of understanding key issues in the distribution of power between risk assessors and risk managers.

In the food safety domain, the sectoral legislation applicable to authorisation procedures for enzymes, additives, flavourings and GM food and feed, provides that the Commission shall justify any divergence from the EFSA opinion (Articles 7(1) and 19(1) of Regulation 1829/2003[60] and Article 7(3) of Regulation 1331/2008[61]). The food safety domain and regulated products in particular have been criticised for the *de facto* regulatory power exercised by EFSA.[62] This trend is not exclusive to food safety, but it has been observed in the aftermath of emerging trends in EU governance related to scientific agencies.[63] EFSA's opinions have been considered substantial—perhaps determinant- on the whole decision-making process.[64] From a *responsibility* perspective, in the light of the active role played by the Authority, it can be deemed to be co-responsible for the decision taken by the risk manager. EFSA oversees what data shall be taken or removed, and how to analyse them. The use of machine learning does not truly change the scenario at stake from an institutional perspective: the relationship between risk assessors and risk managers is the same as before. What changes is the epistemic role played by such algorithms, which now become part of a decision-making process because some tasks are delegated to them.

From an *accountability* point of view, it has been observed that 'as formally advisory agencies have mutated into decision-making ones and old, formal boundaries have been lost, an accountability gap has emerged in the absence of checks on this de facto exercise of decision-making powers'.[65] The introduction of machine learning algorithms is not revolutionary, nor can it mend accountability gaps. However, relying on probabilistic models poses new questions to the alleged lack of accountability mechanisms for EFSA. Without the provision of robust accountability frameworks,

---

[59] Athan et al. (2015)

[60] Regulation (EC) No. 1829/2003 of the European Parliament and of the Council of 22 September 2003 on genetically modified food and feed [2003] OJ L 268/1

[61] Regulation (EC) No. 1331/ 2008 of the European Parliament and of the Council of 16 December 2008 establishing a common authorization procedure for food additives, food enzymes and food flavourings [2008] OJ L 354/1

[62] Kanska (2004)

[63] Dehousse (2002)

[64] Alemanno and Mahieu (2008)

[65] Busuioc (2014)

the Authority could influence decision-making processes by justifying their results on the basis of information generated via inscrutable 'oracles'.

To put it differently, the alleged substantial influence over decision-making processes should not theoretically occur because clear boundaries separate risk assessment from risk management activities. This was established as the post-*Meroni*[66] institutional framework.[67] However, with the progressive erosion of this barrier identified by scholars,[68] EFSA's opinions are progressively becoming substantially binding. An additional layer of influence can be hidden under the alleged accuracy of machine learning algorithms that prevent human oversight or scrutinisability. This conceptualisation of accountability takes into account both the relevance of this principle discussed in the context of food safety and the one proposed in the field of AI ethics.

Liability gaps thus represent an additional challenge. Risk assessors are tasked with scientific evaluations and the Court of Justice of the European Union has confirmed the non-binding nature of EFSA scientific opinions in multiple occasions. In the *Arysta*[69] case, Court's general order noted that 'only measures definitively laying down the position of the institution on the conclusion of that procedure are, in principle, measures against which proceedings for annulment may be brought. It follows that preliminary measures or measures of a purely preparatory nature are not measures against which proceedings for annulment may be brought' (para 43). The exact same wording was used in Court's order in *FMC Chemical v. EFSA*[70] and in Court's order in *Dow Agrosciences v. EFSA*.[71] Only 'manifest error, abuse of powers or clear excess in the bounds of discretion' can fall within the scrutiny of the Court.[72] Therefore, when risk assessors act within their scientific mandate, they cannot be held liable for the decisions taken by risk managers. When addressing the justiciability of the scientific opinions, the CJEU has clearly stated that its review cannot invade the scope of a scientific analysis.[73] Its scrutiny is limited to the procedures followed in the risk analysis, whereas substantial aspects of the risk assessments are left without review.[74]

To summarise, while EFSA is causally responsible, it is not liable for the decision-making process. This situation occurs as the causal contribution to the decision by the risk manager is not translated into a liability regime via some degree of accountability scheme. While solving all the aforementioned concerns

[66] Case 9/56, *Meroni & Co, Industrie Metallurgiche SpA v. High Authority* [1958] ECR 133

[67] Busuioc (2014)

[68] Busuioc (2014) and Dehousse (2002)

[69] Case T-311/06 *FMC Chamical and Arysta Lifesciences v. EFSA* [2008] ECLI:EU:T:2008:205

[70] Case T-312/06 FMC Chemical v. EFSA [2008] ECLI:EU:T:2008:206, para 43

[71] Case T-397/06 *Dow Agrosciences v. EFSA* [2008] ECLI:EU:T:2007:68

[72] Alemanno and Gabbi (2016) and Gabbi (2008)

[73] Case T-13/99, *Pfizer Animal Health SA v. Council* [2002] ECR II-3305, para. 169

[74] Joined Cases T-74/00, 76, 83, 85, 132, 137 and 141, *Artegodan GmbH v. Commission* [2002] ECR II-4945 ()para 200)

might require institutional and governance interventions whose discussion is placed outside the scope of this book, the risks linked to the exacerbation of these concerns due to the use of probabilistic methods can be placed within our discussion. Adequate redress mechanisms the use of machine learning algorithms can mitigate some of the issues highlighted in the previous paragraphs.[75] However, to date, no specific mechanism has been provided in the current legislative framework, not even in the newly enacted 'Transparency Regulation' whose main focus was data-centric. Once again, it is necessary to frame this discussion within the institutional framework under scrutiny. When decisions are taken, the precautionary principle is one of the cornerstones of the decision-making process. Let us see how it interacts with machine learning algorithms.

4.2.3 Machine Learning and the Precautionary Principle

Striking a balance between the responsiveness and the certainty of the analysis is a massive challenge due two clashing interests at stake. While the industry calls for faster 'smart' approval procedures, certainty of the results cannot be achieved without cautious scrutiny and extensive cross-validation tests, which naturally require time to be performed. The precautionary principle, discussed in this section, has been traditionally considered as the guideline to strike this balance. However, the introduction of novel analytics techniques—including machine learning—calls for a discussion on their interaction with the principle.

The precautionary principle has been defined as the necessity that, 'in cases of serious or irreversible threats to the health of humans or ecosystems, acknowledged scientific uncertainty should not be used as a reason to postpone preventive measures'.[76] In data-driven risk assessment, the principle can be understood as the case in which the lack of data originates scientific uncertainty towards a potential threat. When this scenario occurs, preventive measures should be taken by risk managers. *Inter alia*, the European Court of Justice in *Alpharma*[77] framed the relationship between risk assessment and risk management in the light of the precautionary principle when the evaluation revolves around data and forecasts. The Court observed that '[N]otwithstanding the existing scientific uncertainty, the scientific risk assessment must enable the competent public authority to ascertain, on the basis of the best available scientific data and the most recent results of international research, whether matters have gone beyond the level of risk that it deems acceptable for society' (para 175). In a nutshell, the precautionary principle

[75] See the discussion in Quintarelli et al. (2019)

[76] Jasanoff (2016) and Martuzzi et al. (2004)

[77] Case T-70/99 Alpharma Inc. v Council of the European Union ECLI:EU:T:2002:210 [2002] ECR-II-03495

states that whenever scientific uncertainty persists towards some identified hazards, the most cautious decision has to be taken.

In the previous subsection, I have discussed how the causal contribution of EFSA to the final decision is not translated into a liability regime. The precautionary principle is the legal tool through which liability is transferred to the risk manager because it provides guidance to the decision-maker on how to deal with scientific opinions, thus allowing the scrutiny over the final decision. The precautionary principle acts as a constraint to the otherwise unlimited discretion of risk managers. The previous subsection has identified critical remarks towards the lack of proper accountability frameworks in the risk assessor despite its *de facto* regulatory power and stated that machine learning algorithms might reinforce such 'hidden' power by the (perceived) persuasiveness of their results even without scrutinisability or human oversight.

With the introduction of machine learning algorithms, it might be necessary to re-conceptualise how the relationship between risk assessment and management works. When opinions are drafted with the support of (*rectius*, delegation to[78]) some machine learning tools, the presence of this computational approach has to be taken into account. This entails that algorithms shall be considered part of the experimental setting for what concerns the way in which they 'make sense' of data.

This statement has some implications. On the one hand, the risk assessor should be able to critically evaluate the findings gathered using machine learning tools by third parties, including applicants in the fields of regulated productions. On the other hand, if risk managers can be held liable for their decision, the way in which an algorithmic-supported decision is scrutinised has to be reshaped starting from some technical considerations. These questions are posed by the applicability of the precautionary principle and answering them is crucial to mitigate the concerns related to the implications of the deployment of machine learning systems raised above.

Scientific uncertainty at the time of artificial intelligence has a broader meaning. As argued above in Sect. 4.2, one of the key differences between deterministic and probabilistic modelling consists in the presence of an unavoidable statistical error in the outputs generated by the latter computational approach. Therefore, the presence of such statistical error has to be taken into account to reply some open questions: what should the role of confidence scores from a risk manager perspective? Should they be communicated to the risk manager and to the general public? If yes, how? Furthermore, the question of how to define uncertainty when opaque, inexplicable or unexplainable machine learning algorithms is open, similarly to other domains where the Explicability of these algorithms is concerning.

The precautionary principle has to be read in conjunction with the probabilistic nature of machine learning results and careful scrutiny is needed both in classification and regression problems. In the field of emissions into the environment, the CJEU has stated that measurements should be done in 'most unfavourable

[78] I will clarify the difference in Sect. 5.1.1

conditions'.[79] This criterion could be contextualised in the light of the probabilistic nature of machine learning and read jointly with the precautionary principle. Taken together, these elements suggest that unsatisfactory algorithmic performances should trigger the precautionary principle and lead to cautious decisions. This approach implicitly recognises that risk assessors should clearly communicate performance scores to the risk managers and inform the general public accordingly.

4.3 The EU Data Governance Act and its Impact on Food Data Governance

On November 25th, 2020 the EU Commission released a proposal of a Data Governance Act.[80] The Proposal aims to 'facilitate data sharing across the EU and between sectors to create wealth for society, increase control and trust of both citizens and companies regarding their data, and offer an alternative European model to data handling practice of major tech platforms.'[81] The Regulation is aligned with EU fundamental rights and values, and represents the first deliverable of the EU Data Strategy.[82]

If approved, the Data Governance Act will be placed within the 'legislative pentagon' or hexagram[83] with other pieces of legislation that, taken together, shape the EU 'digital constitutionalism'.[84] The Act is also meant to foster trust and legal certainty towards the use of non-personal data.[85] However, early commentators[86] critically noted that such regulatory framework comes at risk of increasing legal certainty and jeopardising compliance efforts.

The section introduces the EU Data Governance Act and its relevance in the food safety data governance. This discussion can be relevant for multiple reasons. With information sharing being desirables among stakeholders, data pooling is hindered by obstacles such as data protection law[87] and intellectual property rights (e.g. in the case of agriculture[88]).

[79] *Bayer CropScience* case para 91. See Sect. 2.2.3 in this book

[80] European Commission Proposal for a Regulation of the European Parliament and of the Council on European data governance (Data Governance Act) COM/2020/767 final

[81] See https://ec.europa.eu/commission/presscorner/detail/en/ip_20_2102, accessed on 27/04/2022

[82] Available at https://digital-strategy.ec.europa.eu/en/policies/strategy-data, accessed on 27/04/2022

[83] Floridi (2021)

[84] De Gregorio (2021)

[85] Kiseleva and De Hert (2021)

[86] Graef and Gellert (2021)

[87] Comandè and Schneider (2021) and Wendehorst (2017)

[88] Sykuta (2016)

4.3.1 The Proposal for an EU Data Governance Act: An Overview

Prior to a food-related discourse, it is necessary to introduce the contents of the proposed Data Governance Act. The general aim of the proposed Regulation is to foster the re-use of public sector data (Recital 3, Recital 4) that falls outside the scope of Directive 2019/1024[89] (Recital 7) and to provide for a 'data altruism' mechanism that allows the use of personal or non-personal data 'without seeking a reward, for purposes of general interest, such as scientific research purposes or improving public services' (Art. 2(10)). To fulfil these goals, providers of data sharing services are established as intermediaries (Chapter II of the Proposed Act). The Act does not aim to grant or modify existing rights over data.[90]

First, definitions are helpful to clarify certain meanings. The definition of 'data' is any 'digital representation of acts, facts or information and any compilation of such acts, facts or information, including in the form of sound, visual or audiovisual recording' (Art. 2(1)). Such definition is sufficiently broad to include personal data (defined in accordance with the GDPR), non-personal data (defined as what is excluded from the definition of personal data) and metadata defined in Article 2(4).[91]

'Re-use' means the use of data held by public institutions by natural or legal persons for purposes other than the original one. This can occur by physical or legal persons, both for commercial and non-commercial purposes (Art. 2(2)).

Then, two subjective legal positions are crucial. On the one hand, the 'data holder' is 'a legal person or data subject who, in accordance with applicable Union or national law, has the right to grant access to or to share certain personal or non-personal data under its control'. Notably, both right holders under the Database Directive and data subjects under the GDPR are granted with a significantly broader *spectrum* of rights, but only two ('access' and 'sharing') are considered by the proposed Data Governance Act to qualify a legal entity as a 'data holder'. European Data Protection Board (EDPB) and the European Data Protection Supervisor (EDPS) have shown some concerns as regards the compatibility of these definitions.[92]

On the other hand, a 'data user' is 'a natural or legal person who has lawful access to certain personal or non-personal data and is authorised to use that data

[89] Directive (EU) 2019/1024 of the European Parliament and of the Council of 20 June 2019 on open data and the re-use of public sector information [2019] OJ L 172/56

[90] Kiseleva and De Hert (2021)

[91] Notably, this definition is consistent with the Communication From the Commission to the European Parliament and the Council 'Guidance on the Regulation on a framework for the free flow of non-personal data in the European Union' COM(2019) 250 final

[92] 'EDPB-EDPS Joint Opinion 03/2021 on the Proposal for a Regulation of the European Parliament and of the Council on European Data Governance (Data Governance Act)' (2021) (pp. 9–11)

for commercial or non-commercial purposes'. The condition of *lawful access* can be controversial when 'data users' can access data publicly available data (hence, lawfully) that are the results of security breaches or intellectual property violation.

'Data sharing' is defined as 'the provision by a data holder of data to a data user for the purpose of joint or individual use of the shared data, based on voluntary agreements, directly or through an intermediary' (Art.2(8)). This definition completes the 'data altruism' introduced above by contemplating the possibility of provisions other than the ones that result in a donation.

As regards the re-use of data held by public sector institutions, Article 3 clarifies that the Regulation would be applicable to data protected on the ground of:

(i) commercial confidentiality;
(ii) statistical confidentiality;
(iii) protection of intellectual property rights of third parties;
(iv) protection of personal data.

When public institutions opt to make this information available for re-use, they cannot enter exclusive or anti-competitive arrangements (Art. 4(1)). Derogations may apply and exclusive right to re-use data can be granted subject to objective conditions (Art. 4(2), Art. 4.3), principles of transparency, equal treatment and non-discrimination (Art. 4(4)), and time limitations (Art. 4(5)).

Competent bodies shall also make publicly available the re-use conditions (Art. 5(1)). Such conditions must be non-discriminatory, proportionate, and objectively justified in the light of: (a) the categories of data, (b) the purposes of re-use, and (c) the nature of the data at stake (Art. 5(2)). They shall not restrict competition. Article 5(3) provides that re-use could be restricted to pre-processed data which may consist of anonymised or pseudonymised personal data or information deprived of commercially-sensitive data. The disclosure of confidential data shall not occur as a consequence of the re-use (Art. 5(8)). Specific rules apply to the transfer of non-personal data to third countries (Art. 5(10–13)), whereas the transfer of personal data is left to the GDPR.

Data sharing services can be categorised in three classes (art. 9):

- with regards to legal persons that qualify as data holders, intermediation services them and potential data users;
- with regards to data subjects, intermediation services between those who seek to make their personal data available and potential data users; and
- data cooperatives, i.e. service providers that support data subjects, one-person companies, micro-enterprises, or SMEs, in the negotiation of the conditions applicable to data sharing in their interest.

Data sharing bodies are subject to a series of obligations (art. 10)—including the notification to a competent authority—and conditions for providing the service (art. 11). They are also subject to compliance monitoring programmes by competent authorities (art. 12 and 13).

Data altruism provisions apply to the free choice of making available personal or non-personal data to third parties without compensation. As clarified above, the donation can occur for purposes of general interest, such as scientific research (to be intended broadly[93]) or to improve public services. Organisations willing to ease data altruism shall register (art. 17) and may operate under the name 'data altruism organisation recognised in the Union' (art. 15). To qualify as such, the organisation shall operate to pursue objectives general interests and on a non-profit scheme (art. 16). Article 19 details the rights and the transparency measures that shall be provided to the data holders (data subjects and legal persons). In particular, the purposes of general interest and the fact the processing takes place outside the EU borders shall be communicated to the data holder. The altruistic organisation shall use the data solely for the purpose of general interest underlying the provision (art. 19(2)). In accordance with Article 14, provisions applicable to data sharing services do not apply to non-profit entities operating in the context of data altruism.

Other provisions include the establishment of a competent authority (Chapter V) and the European Data Innovation Board (Chapter VI).

4.3.2 The Applicability of 'Data Exclusivity' Clauses and Competition Issues

Once identified the main contents of the proposed Data Governance Act for what concerns our focus, let us see how they might interact with the food safety system. This section will explore the first set issue with regards to data exclusivity and competition.

A preliminary question regards food-related data provided by the industry and their falling within the scope of art. 3. The Data Governance Act would be applicable to this data if two conditions had been met: data should be held by a public sector body and they should be protected on the grounds of confidentiality, intellectual property rights rules, or personal data protection laws. Moreover, these data should not fall within the exceptions laid down in Article 3(2). The second condition ('being protected') is relatively easy to meet: the analysis in Chap. 2 reveals that layers of legislation protect personal and non-personal food-related data. The first condition requires an in-depth discussion.

Article 2 of the Proposal states that 'bodies governed by public law' means bodies that have the following characteristics:

(a) they are established for the specific purpose of meeting needs in the general interest, and do not have an industrial or commercial character;
(b) they have legal personality;
(c) they are financed, for the most part, by the State, regional or local authorities, or by other bodies governed by public law; or are subject to management supervision by those authorities or bodies; or have an administrative, managerial

[93] Comandè and Schneider (2021)

or supervisory board, more than half of whose members are appointed by the State, regional or local authorities, or by other bodies governed by public law;

Shall EU agencies—thus, including EFSA—fall within the remit of this provision? Arguments can be raised both for and against the inclusion.

On the one hand, EFSA shall be considered a body governed by public law as it matches conditions (a) and (b). As regards condition (c), EFSA's funding comes from the European Union and 11 out of 15 of its members are elected by the EU Council following consultation with the EU Parliament among a list of experts proposed by the EU Commission (Art. 25 of the GFLR). The Open Data Directive[94] specifies, while mirroring the definition of 'bodies governed by public law' of the Data Governance Act in its Article 2(1–2), that its scope extends to 'existing documents held by public sector bodies *of the Member States* (emphasis added)'. The lack of a similar specification in the Data Governance Act proposal suggests that the Act might be applicable to non-Member States institutions. Article 3(3) establishes that these provisions are 'without prejudice to Union and national law on access to documents and to *obligations of public sector bodies under Union and national law* to allow the re-use of data.' (emphasis added). This statement is quite confusing. It can either be read as 'obligations under Union and national law addressed to public sector bodies' or 'obligations addressed to public sector bodies subject to Union or national law'. In the latter case, EFSA would be considered a public sector body for the purposes of the Data Governance Act, which applies to it without prejudice to other applicable provisions. Finally, a teleological interpretation of the proposed Data Governance Act would recommend not to exclude EU agencies—thus, including EFSA—from the list of bodies subject to this piece of legislation.

On the other hand, our reading of condition (c) of article 2(12) of the Data Governance Act is rather cumbersome and might seem arbitrary: the article refers to funding coming from 'State, regional or local authorities' or 'other bodies governed by public law' defined in the same manner. Justifying that EFSA qualifies as a body governed by public law implies that the EU itself necessarily qualifies as a 'public sector body' for being funded by its Member States. This interpretation cannot be accepted *sic et simpliciter*. Moreover, other provisions in the chapter regulating data re-use (e.g. the competent bodies in Article 7) are addressed to Member States only. Finally, the Data Governance Act is meant to complement the Open Data Directive (Recital 5) in the definition of an EU Data Strategy. The Open Data Directive targets Member States only, thus suggesting that also the Data Governance Act should provide for the same scope.

Pending legislative approval and in the absence of jurisprudence, neither interpretation is fully satisfactory. To circumvent this issue, I will set a cautious *what-if* scenario similar to the one already occurred when EFSA had to deal with the re-use of data and documents of the EU Commission under decision 2011/833/EU.[95]

[94] Directive (EU) 2019/1024 of the European Parliament and of the Council of 20 June 2019 on open data and the re-use of public sector information [2019] OJ L 172/56

EFSA stated as follow: '[a]s a Union agency, EFSA is part of the broader EU policy trend to implement the vision set out in Europe 2020 growth strategy. The European Commission has adopted a Decision regulating use and re-use of its data and documents, which includes a general principle on open data and reusability. Although not directly applicable to the Authority, this important legal act sets the course EFSA should follow. EFSA should implement a presumption of reusability and lay down clearly the conditions when, or processes where, the principle should be validly applied while avoiding unnecessary administrative burdens for interested parties and the Authority itself'.[96]

The hypothetical *what-if* consists of analysing the scenario following the application of the provisions applicable to data re-use under the Data Governance Act, in particular those regarding confidentiality. In Chap. 2, we discussed how EFSA regularly gets access to data provided by applicants willing to place their regulated products on the markets. As argued in the same Chapter, this information is *held* by EFSA. The Authority 'has the right to grant access to or to share certain personal or non-personal data under its control', that is the definition of 'holding' under the Data Governance Act.[97]

Assuming the correctness of the proposed interpretation, food-related data could be subject to the Regulation and to provisions fostering their re-use, without prejudice and in parallel to the application of the Transparency Regulation. Such release can have considerable effects for the competition in the food industry. On the one hand, they could lower R&D costs discussed in Sect. 1.2.3; on the other hand, the competitive position of the original applicant could be hindered. In fact, differently from the Transparency Regulation and its detailed assessment on confidentiality, the Data Governance Act only provides for a few criteria to set the conditions for the re-use.

In particular, Article 5(2) of the Proposal states that 'conditions for re-use shall be non-discriminatory, proportionate and objectively justified with regard to categories of data and purposes of re-use and the nature of the data for which re-use is allowed. These conditions shall not be used to restrict competition'. The possibility offered by Article 5(5) ('Public sector bodies may impose an obligation to re-use only pre-processed data where such pre-processing aims to anonymize or pseudonymise personal data or delete commercially confidential information, including trade secrets') is also applicable to our domain. Finally, Article 5(8) of the

[95] Commission Decision of 12 December 2011 on the reuse of Commission documents (2011/833/EU) [2011] OJ L 330/39

[96] See the presentation available at https://www.efsa.europa.eu/sites/default/files/corporate_publications/files/openefsadiscussionpaper14.pdf, accessed on 27/04/2022

[97] On a side note, the notions of 'data holder' and 'data user' seem quite consistent with the distributed ownership model presented in Sect. 2.3.1 due to their focus on the control over information. However, this wording has not been exempted from critiques regarding its vagueness and its ambiguity when compared to the definitions in the GDPR ('EDPB-EDPS Joint Opinion 03/2021 on the Proposal for a Regulation of the European Parliament and of the Council on European Data Governance (Data Governance Act)' (2021).)

Proposal states that 'when data requested is considered confidential, in accordance with Union or national law on commercial confidentiality, the public sector bodies shall ensure that the confidential information is not disclosed as a result of the re-use.'

Taken together, these provisions set criteria to draft re-use conditions:

- The 'categories of data' (art. 5(2)) to be taken into account are mainly non-personal. The qualification of this data under the remit of environmental information has to be preliminarily assessed to verify the joint application of the Regulations that mandate the disclosure. Nonetheless, as argued when discussing the processing of food consumption data, it might be the case that re-identifiable personal data processed by applicants are released for re-use. This calls for an assessment as regard the degree of aggregation/anonymisation similarly to the one discussed in Chap. 3. Concerns about the categorisation of food related data discussed in Chap. 3 as regards the applicability of data protection law might be overcome by the horizontal application of the Data Governance Act to personal and non-personal data. Instead, as noted by the EDPB-EDPS joint analysis,[98] data falling under the Data Governance Act might be temporarily subject to the GDPR and to Regulation 2018/1807 on the free flow of non-personal data;
- The 'purposes of re-use' (art. 5(2)) can range from the cross-validation of scientific results from the entering into the market as a competitor of the applicant. In the latter case, stricter re-use conditions might apply to preserve the competitive position of the applicant;
- The 'nature of the data' (art. 5(2)) criterion seems overlapping with the 'categories of data' criterion. It might be interpreted broadly to include a verification on the level of aggregation, the presence of intellectual property rights, and other elements which fall outside data protection law categories;
- The use of 'pre-processed data' (art. 5(5)) is a viable alternative to the re-use of raw information. In particular, the deletion of 'commercially confidential information' is compatible with the release of the non-confidential version of the data provided by the industry provided by the Transparency Regulation;
- The preservation of confidentiality (art. 5(8)) following the re-use is perhaps the hardest requirement to be met by EFSA when releasing data for re-use. In particular, when the release is meant to foster the cross-validation of scientific result, a careful balance has to be struck when setting re-use conditions. They should allow for sufficient data analysis for the cross-check without hindering the confidentiality agreed to the applicant. Despite its vagueness in the absence of technical specifications, the 'secure processing environment' provided by Article 5(4)(a) seem a possible implementation.

These criteria seem compatible with the release mechanisms provided by the Transparency Regulation. However, the Data Governance Act represents a step

[98] 'EDPB-EDPS Joint Opinion 03/2021 on the Proposal for a Regulation of the European Parliament and of the Council on European Data Governance (Data Governance Act)' (2021) p. 7

forward the possibility of release conditions that pave the way for data re-use. Beside speculations, the full assessment of compatibility can occur only after the approval of the Regulation.

4.3.3 Data Altruism, Data Cooperatives, General and Public Interest

Other provisions of the proposed Data Governance Act seem even more compatible with the current legislative framework in food safety. In comparison to data re-use provisions, the advantage of these rules is that they do not require arguable interpretative attempts to include EFSA within their remit. Let us analyse some of these provisions.

First, 'data altruism' is seemingly overlapping with the common practice of 'calls for data'.[99] Such calls allow EFSA to access information of third party who are voluntarily called to participate risk assessment. Calls can be referred to technical data, toxicological information, and literature searches on a given topic. Depending on the submitted information, confidentiality treatment might be requested. Together with this consolidated procedure, data altruism organisation—such as NGOs—could register to ease such transfer.

Doubts regarding the permission to re-use have been posed in a 2009 paper.[100] If entered into force, the Data Governance Act would allow re-use insofar it is compatible with the original purpose (art. 19(2)) and information regarding these purposes have to be provided to the data contributor.

Instead, other voluntary contributions can be explored outside 'calls for data'. In particular, individuals could decide to share their food consumption data.[101] The Data Governance Act could foster such forms of collaboration via qualified intermediaries. One possible source of concern regards the general interest that shall be pursued by the data receiver. Given the possibility of re-using dietary intake information for purposes other than food safety risk assessment due to the high inferential power of these data, it is necessary to set up measures that effectively limit the power to dispose of this information. Notably, the EDPB-EDPS joint analysis has highlighted some inconsistencies between the notion of 'permission' for the data sharing and the legal basis for consent under the GDPR.[102]

Likewise, data cooperatives (one possible form of providers of data sharing services) can be explored in farming. Commentators from an NGO[103] have argued about the lack of clarity in the ownership of farming data within the Data

[99] See https://www.efsa.europa.eu/en/calls/data?page=1, accessed on 27/04/2022

[100] Kocharov (2009)

[101] As it has been observed, individuals have more access to their dietary intake information thanks to smartphones and 'smart' devices in comparison to the past

[102] 'EDPB-EDPS Joint Opinion 03/2021 on the Proposal for a Regulation of the European Parliament and of the Council on European Data Governance (Data Governance Act)' (2021), p. 13

Governance Act. The issue is raised by the lack of new rights to collective data management that would entitle farmers to share data produced by their machineries. Therefore, it is first necessary to allow farmers legitimately dispose of these data for these purposes. This can be done by multiple means, including specific provisions in contractual agreements that give access and ownership rights to the farmers,[104] or other interested processors along the food chain. To put it differently, farmers should first qualify as 'data holders' according to the definition of the Data Governance Act before proceeding with the altruistic sharing. As with the case of dietary intake information, farming data should not be used for other purposes—such as price discrimination—which might negatively adverse them and lower their trust, which is necessary to foster data sharing.[105] This seems consistent with the *desideratum* of authors arguing in favour of a public or societal interests to be balanced with those of originators[106] and, in general, with the aims of the European Union Code of Conduct on Agricultural Data Sharing by Contractual Agreement promoting digitally-enhanced farming.

A possible question regards the level of detail of the original purpose. For instance, risk assessment analysis range from outbreak monitoring to pesticide approval: while an individual, a farmer or a cooperative would be happy to contribute to the monitoring of food-related illnesses, it might be that case that the same entity would not contribute to pesticide approval procedures. In other words, the purpose of the altruistic contribution should be detailed enough to prevent misuses, while sufficiently broad to allow some degree of re-use. One significant limitation that also affects food safety illustrated by some commentators[107] consists of the lack of mention of mixed datasets, i.e. 'the majority of datasets available in the data economy' according to the EU Commission.[108]

4.4 Chapter Synopsis

This Chapter has presented and discussed ongoing trends in food safety data governance that raise questions that are left unanswered by the legal framework at stake. This chapter signifies a shift in this book from known issues—data ownership

[103] See https://itforchange.net/sites/default/files/add/ITfC%27s-input-to-draft-DGA.pdf, accessed on 27/04/2022

[104] See also the European Union Code of Conduct on Agricultural Data Sharing by Contractual Agreement at http://www.fao.org/family-farming/detail/en/c/1370911/, accessed on 27/04/2022

[105] Burg et al. (2020)

[106] Kiseleva and De Hert (2021)

[107] Graef and Gellert (2021)

[108] Communication From the Commission to the European Parliament and the Council 'Guidance on the Regulation on a framework for the free flow of non-personal data in the European Union' COM(2019) 250 final

and personal data processing—to a relatively unknown field of study in food safety that is represented by the combination of some elements.

First, the technicalities underlying the shift from deterministic to stochastic analysis have been presented. In summary, certain emerging concerns are identified. The possibility of biases in data analysis is introduced, in particular as regards minoritarian food patterns and literature reviews, two intertwined issues that deserve attention. Then, the 'black box' issue and the inscrutability of the logic underlying certain outputs is discussed and tackled with a Knowability approach that suggests the disclosure of data, algorithms, and operational information as a solution to mitigate the 'black box' concern.

One set of issues is related to the probabilistic nature of algorithmic outputs. On the one hand, it is possible that results are biased and polarised, with discriminatory outcomes as regards certain groups; on the other hand, correctly working software can generate probabilistic failures due to the nature of their results (e.g. false positives/true negatives). These concerns generate in turn the problem of how to assess *unfair* results. The Chapter then discusses what is fairness in our domain and how it interacts with data analysis and machine learning. Allocating the responsibility of unfair decisions based on algorithms is thus necessary. The Chapter has presented a threefold explanation of the responsibility, accountability and liability schemes in our research scenario: taken together, causal responsibility is disentangled from the attribution of moral accountability and legal liability. While the first is mainly contributed by EFSA, the Authority is not liable as the judicial scrutiny is limited to the risk managers. The precautionary principle is then assumed as a viable redress mechanism. It governs the assessment/management relationship and provide guidance on how to handle scientific uncertainty. If interpreted in the light of probabilistic modelling (Sects. 5.2.2 and 5.2.4), it can prevent probabilistic failures and serve as a criterion to allocate the responsibilities among the involved institutions.

Finally, the proposed Data Governance Act is introduced and discussed as a forthcoming piece of legislation. Despite the obvious limitations due to the ongoing legislative debate, the proposed Data Governance Act seems promising as regards two sets of provisions: on the one hand, limitations in 'data exclusivity' clauses and open possibilities for data re-use pave the way for additional tools to further data disclosures; on the other hand, data altruism and free contributions of data can broaden the scope of EFSA's calls for data and provide guidance for altruistic contributions of data by parties such as individuals of cooperatives.

References

AGES, Vlachou C, Hofstädter D (2019) Joint venture on the further development of chemical exposure assessment by use of probabilistic modelling. EFSA J 17:e170905

AI4People (2018). AI4People | Atomium. https://www.eismd.eu/ai4people/

Alemanno A, Gabbi S (2016) Foundations of EU food law and policy: Ten years of the European food safety authority. Routledge, London

Alemanno A, Mahieu S (2008) The European food safety authority before European courts: Some reflections on the judicial review of EFSA scientific opinions and administrative acts. Eur Food Feed Law Rev, 320–333

Arrieta AB, Diaz-Rodriguez N, Del Ser J, Bennetot A, Tabik S, Barbado A, Garcia S, Gil-Lopez S, Molina D, Benjamins R et al. (2020) Explainable artificial intelligence (XAI): Concepts, taxonomies, opportunities and challenges toward responsible AI. Inform Fusion 58:82–115

Athan T, Governatori G, Palmirani M, Paschke A, Wyner A (2015) LegalRuleML: Design principles and foundations. In: Reasoning web international summer school. Springer, Berlin, pp 151–188

Barabas C (2020) Beyond bias: Re-imagining the terms of "ethical AI" in criminal law. Geo J Law Mod Critical Race Persp 12:83

Barocas S, Selbst AD (2016) Big data's disparate impact. Calif Law Rev 104:671

Bellamy RKE, Dey K, Hind M, Hoffman SC, Houde S, Kannan K, Lohia P, Martino J, Mehta S, Mojsilović A et al. (2019). AI Fairness 360: An extensible toolkit for detecting and mitigating algorithmic bias. IBM J Res Develop 63(4/5):4–1

Bisgin H, Bera T, Ding H, Semey HG, Wu L, Liu Z, Barnes AE, Langley DA, Pava-Ripoll M, Vyas HJ et al. (2018) Comparing SVM and ANN based machine learning methods for species identification of food contaminating beetles. Sci Rep 8(1):1–12

van der Burg S, Wiseman L, Krkeljas J (2020) Trust in farm data sharing: reflections on the EU code of conduct for agricultural data sharing. Ethics Inform Technol, 1–14

Busuioc EM (2014) Blurred areas of responsibility: European agencies' scientific 'opinions' under scrutiny. In: Monika A, Karin A, Ellen H, Helena R (eds) The role of 'experts' in international and European decision-making processes: Advisors, decision makers or irrelevant actors?. Cambridge University Press, Cambridge, pp 383–402. https://doi.org/10.1017/CBO9781139871365.023

Castañón, CAB, Fraga JS, Fernandez S, Gruber A, da F Costa L (2007) Biological shape characterization for automatic image recognition and diagnosis of protozoan parasites of the genus Eimeria. Pattern Recogn 40(7):1899–1910

Comandè G, Schneider G (2021) Can the GDPR make data flow for research easier? Yes it can, by differentiating! A careful reading of the GDPR shows how EU data protection law leaves open some significant flexibilities for data protection-sound research activities. Comput Law Secur Rev 41

Contissa G, Laukyte M, Sartor G, Schebesta H, Masutti A, Lanzi P, Marti P, Tomasello P (2013). Liability and automation: Issues and challenges for socio-technical systems. J Aerosp Oper 2(1–2):79–98

de Fine Licht K, and de Fine Licht J (2020) Artificial intelligence, transparency, and public decision-making. decision-making. AI Soc 1–10

De Gregorio G (2021) The rise of digital constitutionalism in the European Union. Int J Constitutional Law 19 (1): 41–70.

De Minico G (2019) Big Data e la debole resistenza delle categorie giuridiche. Privacy e lex mercatoria. Diritto pubblico 25(1):89–116

Dehousse R (2002) Misfits: EU law and the transformation of European governance, in Christian Joerges, and Renaud Dehousse (eds), Good Governance in Europe's Integrated Market, Collected Courses of the Academy of European Law (Oxford, 2002; online edn, Oxford Academic, 22 Mar. 2012)

Durante M (2017) Ethics, law and the politics of information: A guide to the philosophy of Luciano Floridi. The international library of ethics, law and technology. Springer, Amsterdam. ISBB: 9789402411508

Durante M (2019) Potere computazionale: L?impatto delle ICT su diritto, società, sapere. Mimesis, Paju-si.

EFSA (2018) Use of EFSA pesticide residue intake model (EFSA PRIMo revision 3). EFSA J 16(1):e05147

EFSA, Dujardin B, Bocca V (2019) Cumulative dietary exposure assessment of pesticides that have chronic effects on the thyroid using SAS®software. EFSA J 17(9):e05763

European Commission (2020) Commission White Paper on Artificial Intelligence - A European approach to excellence and trust. https://ec.europa.eu/info/sites/info/files/commission-white-paper-artificial-intelligence-feb2020_en.pdf. Accessed 26 April 2022

Ferentinos KP, Yialouris CP, Blouchos P, Moschopoulou G, Kintzios S (2013) Pesticide residue screening using a novel artificial neural network combined with a bioelectric cellular biosensor. BioMed Res Int 2013

Floridi L (2013) The ethics of information. Oxford University Press, Oxford

Floridi L (2021) The European Legislation on AI: a Brief Analysis of its Philosophical Approach. Philos Technol 1–8

Floridi L, Cowls J, Beltrametti M, Chatila R, Chazerand P, Dignum V, Luetge C, Madelin R, Pagallo U, Rossi F et al. (2018) AI4People—An ethical framework for a good AI society: opportunities, risks, principles, and recommendations. Minds Mach 28(4):689–707

Gabbi S (2008) The European Food SafetyAuthority: judicial reviewby community courts. Revue européenne de droit de la consommation (REDC). Eur J Consumer Law 1(2009):171–189

Graef I, Gellert R (2021) The European Commission's proposed Data Governance Act: some initial reflections on the increasingly complex EU regulatory puzzle of stimulating data sharing. In P. T. J. Wolters, R. M. Hermans, A. U. Janssen, & P. Ortolani (Eds.), Digitalisering en conflictoplossing (pp. 219–240)

Gu W, Vieira AR, Hoekstra RM, Griffin PM, Cole D (2015) Use of random forest to estimate population attributable fractions from a case-control study of Salmonella enterica serotype Enteritidis infections. Epidemiol Infect 143(13):2786–2794

Hacker P (2018) Teaching fairness to artificial intelligence: Existing and novel strategies against algorithmic discrimination under EU law' 55, Common Market Law Review, Issue 4, pp. 1143–1185

Hallevy G (2010) The criminal liability of artificial intelligence entities-from science fiction to legal social control. Akron Intell Prop J 4:171

Hallinan D, de Hert P (2017) Genetic classes and genetic categories: protecting genetic groups through data protection law. In: Group privacy. Springer, Berlin, pp 175–196

High-Level Expert Group on Artificial Intelligence, HLEG (2019) Ethics Guidelines for Trustworthy AI. Available at https://digital-strategy.ec.europa.eu/en/library/ethics-guidelines-trustworthy-ai. Accessed on 27/04/2022

IZSTO, Ru G, Crescio MI, Ingravalle F, Maurella C, UBESP, Gregori D, Lanera C, Azzolina D, Lorenzoni G et al. (2017) Machine Learning Techniques applied in risk assessment related to food safety. EFSA Support Public 14(7):1254E

Jasanoff S (2016) The ethics of invention: Technology and the human future. WW Norton & Company, NewYork

Kanska K (2004) Wolves in the clothing of sheep? The case of the European Food Safety Authority. Eur Law Rev 5:711–727

Kiseleva A, De Hert P (2021) Creating a European health data space. Obstacles in four key legal areas. European Pharmaceutical Law Review (EPLR). Lexxion Publisher, Berlin, Forthcoming.

Klaveren JD van, Kruisselbrink JW, de Boer WJ, van Donkersgoed G, te Biesebeek JD, Sam M, van der Voet H (2019) Cumulative dietary exposure assessment of pesticides that have chronic effects on the thyroid using MCRA software. EFSA Support Public 16(9):1707E

Kocharov A (2009) Data ownership and access rights in the European food safety authority. Eur Food Feed Law Rev 335–346

Lagioia F, Contissa G (2020) The strange case of Dr Watson: liability implications of AI evidence-based decision support systems in health care. Eur J Legal Stud 12:245

Lambe J (2002) The use of food consumption data in assessments of exposure to food chemicals including the application of probabilistic modelling. Proc Nutr Soc 61(1):11–18

Malgieri G (2020) The concept of fairness in the GDPR: a linguistic and contextual interpretation. In: Proceedings of the 2020 conference on fairness, accountability, and transparency, pp. 154–166

Martuzzi M, Tickner JA et al. (2004) The precautionary principle: protecting public health, the environment and the future of our children. World Health Organization, Geneva
Micklitz H-W, Pollicino O, Reichman A, Simoncini A, Sartor G, De Gregorio G (2021) Constitutional challenges in the algorithmic society. Cambridge University Press, Cambridge
Miller T (2019) Explanation in artificial intelligence: Insights from the social sciences. Artif Intell 267:1–38
Mittelstadt Br, Floridi L (2016) The ethics of big data: Current and foreseeable issues in biomedical contexts. Sci Eng Ethics 22(2):303–341
Mittelstadt B, Russell C, Wachter S (2019) Explaining explanations in AI. In: Proceedings of the conference on cairness, accountability, and transparency, pp 279–288
Nalepa G, Araszkiewicz M (2020) Liability and responsibility of intelligent systems [in eng]. In: Maria Jesus G-E, Pavon J (eds) An introductory guide to artificial intelligence for legal professionals. Wolters Kluwer, Alphen aan den Rijn, pp 109–126. ISBN: 978-94-035-0943-3
Ortiz-Pelaez Á, Pfeiffer DU (2008) Use of data mining techniques to investigate disease risk classification as a proxy for compromised biosecurity of cattle herds in Wales. BMC Vet Res 4(1):24
Otterbacher J (2018) Addressing social bias in information retrieval. In: International conference of the cross-language evaluation forum for European languages. Springer, Berlin, pp 121–127
Pagallo U (2017) From automation to autonomous systems: A legal phenomenology with problems of accountability. In: 26th international joint conference on artificial intelligence, IJCAI 2017, pp 17–23
Pagallo U (2020) Algoritmi e conoscibilità. Rivista di Filosofia del Diritto 9(1):93–106
Pajno A, Bassini M, De Gregorio G, Macchia M, Patti FP, Pollicino O, Quattrocolo S, Simeoli D, Sirena P (2019) AI: profili giuridici. Intelligenza Artificiale: criticità emergenti e sfide per il giurista. BioLaw J Rivista di BioDiritto 18(3):205–235
Palmirani M (2020) Big Data e conoscenza. Rivista di Filosofia del Diritto 9(1):73–92
Palmirani M, Sapienza S (2021) Big data, explanations and knowability ragion pratica. Ragion Pratica (2), 349–364
Papakyriakopoulos O, Hegelich S, Serrano JCM, Marco F (2020) Bias in word embeddings. In: Proceedings of the 2020 conference on fairness, accountability, and transparency, pp 446–457
Pasquale F (2015) The black box society. Harvard University Press, Cambridge
Quintarelli S, Corea F, Fossa F, Loreggia A, Sapienza S (2019) Una prospettiva etica sull'Intelligenza Artificiale: principi, diritti e raccomandazioni. Rivista di Biodiritto 3:183–204
Ribeiro MT, Singh S, Guestrin C (2016a) Why should i trust you? Explaining the predictions of any classifier. In: Proceedings of the 22nd ACM SIGKDD international conference on knowledge discovery and data mining, pp 1135–1144
Ribeiro MT, Singh S, Guestrin C (2016b) Model-agnostic interpretability of machine learning. Preprint arXiv:1606.05386
Russell S, Norvig P (2010) Artificial intelligence: a modern approach. Pearson, London
Sovrano F, Vitali F, Palmirani M (2019) The difference between Explainable and Explaining: requirements and challenges under the GDPR. XAILA @JURIX 2019. Available at http://ceur-ws.org/Vol-2681/xaila2019-paper1.pdf
Sovrano F, Sapienza S, Palmirani M, Vitali F (2022) Metrics, explainability and the European AI act proposal. Multidiscip Sci J 5(1):126–138
Strawn LK, Fortes ED, Bihn EA, Nightingale KK, Gröhn YT, Worobo RW, Wiedmann M, Bergholz PW (2013) Landscape and meteorological factors affecting prevalence of three food-borne pathogens in fruit and vegetable farms. Appl Environ Microbiol 79(2):588–600
Sykuta ME (2016) Big data in agriculture: property rights, privacy and competition in ag data services. Int Food Agribus Manag Rev 19(1030-2016-83141):57–74
Tsamados A, Aggarwal N, Cowls J, Morley J, Roberts H, Taddeo M, Floridi L (2022) The ethics of algorithms: key problems and solutions. AI Soc 37(1):215–230
Veale M, Van Kleek M, Binns R (2018) Fairness and accountability design needs for algorithmic support in high-stakes public sector decision-making. In: Proceedings of the 2018 chi conference on human factors in computing systems, pp 1–14

Wachter S, Mittelstadt B, Floridi L (2017a) Transparent, explainable, and accountable AI for robotics. Sci Robot 2(6)

Wachter S, Mittelstadt B, Floridi L (2017b) Why a right to explanation of automated decision-making does not exist in the general data protection regulation. Int Data Privacy Law 7(2):76–99

Wendehorst C (2017) Elephants in the room and paper tigers: How to reconcile data protection and the data economy. In: Trading data in the digital economy: Legal concepts and tools. Nomos Verlagsgesellschaft mbH & Co. KG, Baden-Baden, pp 327–356

Zarsky T (2016) The trouble with algorithmic decisions: An analytic road map to examine efficiency and fairness in automated and opaque decision making. Sci Technol Human Values 41(1):118–132

Zerilli J, Knott A, Maclaurin J, Gavaghan C (2019) Transparency in algorithmic and human decision-making: Is there a double standard? Philos Technol 32(4):661–683

Chapter 5
The P-SAFETY Model: A Unifying Ethical Approach

5.1 Towards an Ethical Blueprint

This chapter finally presents the core proposal of this book, i.e., an ethical Blueprint intended to contextualise high-level AI principles to the domain of food safety. The Blueprint is both a proposal and a summary of our findings so far. As discussed in the Introduction, our task was to adapt ethical principles in AI to a relatively unknown research domain. This task has a twofold implication: on the one hand, it contributes to apply such principles to practical scenarios where they can be effective; on the other hand, the resilience of these principles is tested against a real-life domain that contributes to the debate on AI governance.

I have to preliminary restate that the Blueprint covers both data and algorithms. While this shift might seem confusing for some, algorithms—including machine learning approaches—are the way in which we 'make sense' of data, understand our datafied reality and take decisions. Restraining our Blueprint to data would simply neglect that, in practice, issues arise for data and algorithms, hence the necessity of holistic solutions.

To introduce, present and discuss the Blueprint, this chapter is structured in the following way:

Section 5.1 introduces the goals and some preliminary remarks on how the Blueprint has been drafted. In particular, it takes into account the role of trust and regulatory approaches to AI. Then it summarises the contents of some "ethical charters" that constitute the conceptual background of the Blueprint.

Section 5.2 presents the contents of the Blueprint and the P-SAFETY (Privacy, Security, Accountability, Fairness, Explainability, Transparency) model, which is its core. Each principle is defined, contextualised within the current academic debate, and applied to the domain of food safety.

Section 5.3 discusses possible implementations of the P-SAFETY model under governance framework, 'hard' or 'soft' laws, and jurisprudence.

S. Sapienza, *Big Data, Algorithms and Food Safety*, Law, Governance and Technology Series 52, https://doi.org/10.1007/978-3-031-09367-8_5

5.1.1 *Trust-Centric Ethics of Artificial Intelligence*

Among its characteristics, the proposed Blueprint is meant to promote trust. Understanding the relevance of trust in this research domain is a crucial passage of my line of reasoning. Therefore, I will devote this subsection to this task.

First, trust has a twofold importance in our scenario. While seemingly disentangled, trust in the food safety system and in the artificial intelligence systems have some common traits that have to be explored. As regards the first domain, it is worth reminding that trust is one of the cornerstones of the food safety system (Recital 9 of the GFLR). Elsewhere, I have also discussed how a low level of trust in the food safety system can revolve around technological issues, in particular poor data quality and altered information.[1] Accuracy in food safety risk assessment is not the only source of concern: openness and transparency of the information used in the course of scientific analyses are deemed to increase trust.[2]

Trust in digital technologies has become the subject of academic research when machines became used to perform intellectual works[3], thus developing a human-to-artificial agent framework of trust[4]. Durante[5] contextualised within an informational environment and according to a socio-cognitive model of trust: it is rationally possible to trust only in relation to a given goal. Trust thus consists of "a stratified dimension made up of: a trust attitude, a decision to trust, and an act of trusting":

- The mental state of the trustor (i.e., the attitude) corresponds to positive expectations about the performance of the task and the relevance of external factors in trustee's performance and can be heuristically quantified as a scalar measure.
- The decision to trust is then represented by the establishment of a communication that shall clarify at minimum the degrees of interactivity, autonomy, and adaptability of the system.
- The act of trusting is only possible if there is an agreement between the trustor and the trustee as regards key aspect of their cooperation, including the knowledge of the structure of the task, the goal to be achieved, and shared concerns about achieving it.

Trust is characterised by a balance between control and uncertainty (or lack of control). Judgement on this balance is further complicated as our evaluations of these dimensions are different by nature.[6]

[1] See Sect. 2.1.3

[2] See Sect. 2.2.3

[3] Floridi (2008) and Taddeo (2010)

[4] Grodzinsky et al. (2011)

[5] Durante (2010)

[6] See the example in Durante (2010) at p. 357

Such model of trust is applicable to autonomous agents.[7] However, as argued, the low degree of autonomy of machine learning systems deployed in our domain compared to others calls for an adaptation of this conceptual framework.[8] With less autonomy on the agents at stake, it might be the case that our machine learning algorithm works in predictable ways, thus avoiding the necessity of trusting it.[9] However, it has been discussed how unpredictable can certain machine learning outputs be in our domain, in particular due to the possibility of biases[10] and opaqueness of their inner logic.[11] Predictability is a key requirement of trust in artificial agents[12] and concerns the adaptation to new and unprecedented cases, beside autonomy. In our domain, algorithms are *adaptive*—rather than autonomous—in their interaction with a given informational environment.

In recent years, trustworthiness of AI systems has been discussed in the light of the growing 'AI for social good' research interest,[13] which revolves around having a positive impact on human life and environment by means of AI. Trustworthiness has been deemed essential for the promotion of positive AI systems.[14] Food safety can be safely placed within the UN SDGs thanks to its contribution to human growth and sustainability.[15] Scholars have argued that "falsifiability is essential factor to improve the trustworthiness of technological applications in general, and AI4SG applications in particular".[16] In this discourse, falsifiability is described as the possibility of specifying and testing critical requirements for the AI, such as its safety.

Once clarified the twofold relevance of trust and trustworthiness, it is possible to discuss how this requirement affects our domain. The picture below describes two delegation processes. Humans and animals do not assess the safety of their foods in ways other than some intuitive (naïve) controls made using their senses and instinct (for instance, unexpected or bad smell sometimes suggests that food is not edible). The "hard tasks" of risk assessment—i.e., those that present a medium level of complexity—are delegated to institutional risk assessors which dispose of the capability and the instruments to perform the risk assessment. This create some degree of information asymmetry between consumers and the other entities involved in the process (independent scientists, food business operators, authorities, ecc.). In turn, risk assessors delegate to machine learning algorithms the "harder tasks" that involve large, mutable and heterogeneous data, thus requiring

[7] Durante (2010), Taddeo (2010), and Mittelstadt et al. (2016)

[8] See Sects. 1.1.3 and 4.2.1

[9] Durante (2010)

[10] See Sect. 4.2.2

[11] See Sect. 4.1.3

[12] Grodzinsky et al. (2011)

[13] Taddeo and Floridi (2018)

[14] Cowls et al. (2019)

[15] See Sect. 1.1.1

[16] Cowls et al. (2019)

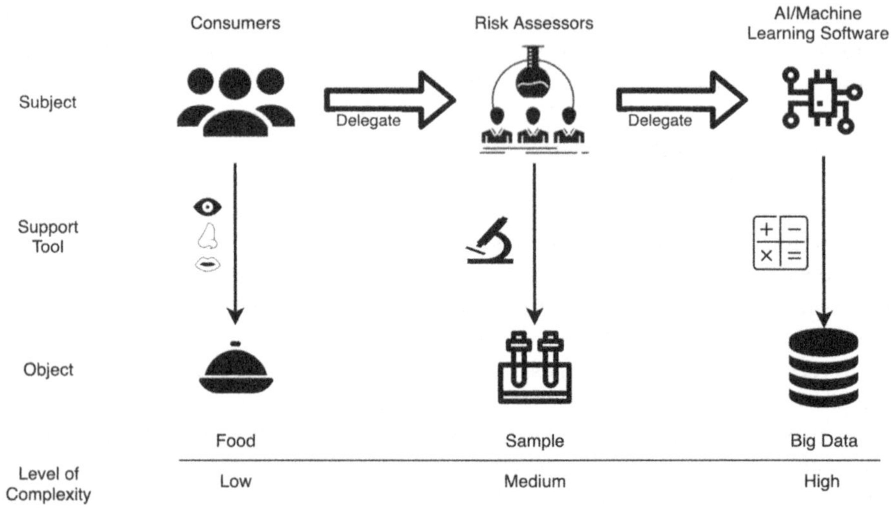

Fig. 5.1 Delegating delegation. Each arrow represents a delegation to another entity of some risk assessment task

higher-level computations. As a result, an additional layer of complexity and information asymmetry from the consumers' perspective is generated. In this scenario, consumers confront with a *delegating delegation* situation represented by Fig. 5.1.[17]

In the course of delegation processes, trust becomes relevant due to information asymmetry between consumers and the food safety community. As argued above and as shown in the picture below, trust is a key element of information collection, analysis and sharing among the entities involved in the food safety system (*first-order* trust). Then, the trustworthiness of AI/machine learning systems is a necessary requirement for the delegation of certain risk assessment tasks to algorithms (*second-order* trust).

Is it also possible to push further these trust-based relationships and infer the existence of an additional form of trust, i.e., the one that links consumers to AI/machine learning algorithms (third-order trust)? A tentative positive answer is suggested. It can be described as the combination of the previous orders: from the first-order trust, the third level embeds the concerns towards the quality and the accuracy of risk assessment activities, including data processing; from the second-order trust, the third level embeds the need of scientific requirements such as reliability, robustness, and falsifiability of the AI systems. To put it differently,

[17] The figure represents a simplified model delegations in food safety risk assessments. As one of the reviewers has correctly pointed out, the presence of other entities—including individual experts, expert groups, AI systems' developers and users, data originators, etc.—should be also taken into account in a possible refined model

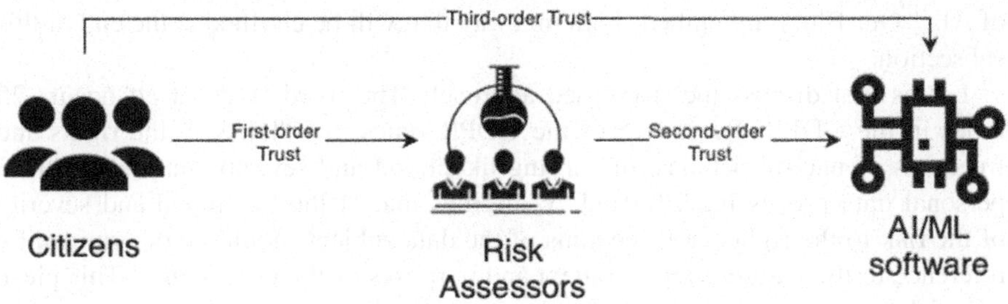

Fig. 5.2 First, second and third-order trust. Each arrow represents a Trustor (nock) -Trustee (tip) relationship

third-order trust corresponds to the general acceptability of AI systems used in the food safety system, yet mediated by an intermediary. Risk assessor thus guarantees the fairness and the reliability of the results in accordance with the responsibility/accountability/liability framework,[18] while ensuring that results are knowable to mitigate information asymmetry.[19]

To summarise, while first-order trust is not new, as food safety authorities pre-existed the current datafication process, second-order and third-order trust are emerging phenomena that correlate with the deployment of machine learning systems. The Blueprint seeks to answer to promote every order of trust (Fig. 5.2).

5.1.2 Risk-Based and Principle-Based Approaches to AI Regulation

Regulating emerging technologies is a rather complex task. New techniques often bring social changes and legislative efforts have to be made to steer new technologies towards acceptable results and social good. AI and machine learning systems are not exceptional in this sense.[20] This subsection identifies two possible legislative approaches that have been adopted by policymakers in the regulation of AI and machine learning techniques.

It has to be preliminary noted that our classification has a purely descriptive function and describes two components of legislative efforts that are intertwined. In particular, we will refer to sources in the 'black letter' law regulatory framework

[18] See Sect. 4.3

[19] See the Knowability approach discussed in Sect. 4.3.1 and explainability requirements set forth in Sect. 5.2.4

[20] Cowls et al. (2019)

of AI.[21] Our Blueprint embeds both in a way that will be clarified at the end of this subsection.

Let us first discuss the risk-based approach. The word "risk" itself occurs 75 times in the GDPR. Recital 75 of the GDPR states that "[r]isk to the rights and freedoms of natural persons, of varying likelihood and severity, may result from personal data processing". Recital 76 clarifies that "[t]he likelihood and severity of the risk to the rights and freedoms of the data subject should be determined by reference to the nature, scope, context and purposes of the processing." This piece of legislation also provides specific rules for processing which might results in risks for fundamental rights and freedoms (e.g., the Data Protection Impact Assessment (DPIA)) in art. 35.

The risk-based approach is inherently consequentialist. Following an assessment of risks linked to a given technology, regulators adopt a normative framework aimed at minimising them. However, commentators have observed that in the GDPR "risk is composed not only of impacts (or consequences, also known as harms when they are negative), but also of events",[22] such as non-compliance. In fact, Recital 90 refers to "risk sources" (rather than "risk factors"). Risk mitigation is then performed by technical and organisational measures (data protection *by design*), which include pseudonymisation, encryption and other technical tools.

The proposal for an EU Artificial Intelligence Act[23] follows a risk-based approach to the regulation of AI. The Commission's proposal identifies different risk levels[24] related to the use of AI systems, which we adapt to analyse in four categories:

- Unacceptable risks, which entails to prohibited systems (Title II);
- High risks systems for 'health, safety and fundamental rights', which entail to a list of applications subject to stricter rules (Title III);
- limited risks, which entail only transparency obligations (Title IV);
- minimal risks, which can be subject to code of conducts (Title IX).

Recital 43 is key to understand the risk-based approach of this Proposal. It states that "[r]equirements should apply to high-risk AI systems as regards the quality of data sets used, technical documentation and record-keeping, transparency and the provision of information to users, human oversight, and robustness, accuracy and cybersecurity. Those requirements are necessary to effectively mitigate the risks for health, safety and fundamental rights as applicable in the light of the intended

[21] Petit and De Cooman (2021)

[22] Gellert (2018)

[23] Proposal for a Regulation of the European Parliament and of the Council laying down harmonised rules on Artificial Intelligence (Artificial Intelligence Act) and amending certain union legislative acts COM(2021) 206 Final

[24] This structure is similar to the one originally proposed by the German Data Ethics Commission "Opinion of the Data Ethics Commission (Executive Summary)" (October 2019)

purpose of the system, and no other less trade restrictive measures are reasonably available, thus avoiding unjustified restrictions to trade".

In its essence, the risk-based approach proposed by the GDPR and endorsed by the proposal for an AI Act can be considered the result of a balance based on the necessity of preserving fundamental rights and freedoms while ensuring the development of AI industry or AI-based products.

Notably, this approach is very popular in the field of food-related regulated products.[25] As we have observed, EFSA is in charge of assessing the safety of potential hazardous products before their placement on the market. In the AI Act and in the GDPR, risk evaluation is performed by the data controller or the AI provider via a risk management system (Art. 9). As noted by early commentator the Proposal is inspired by product safety regulatory frameworks.[26]

On the other hand, the principle-based approach to regulation stresses the promotion of ethical principles when new technologies are developed. For instance, the GDPR also prescribes principles in Art. 5. In fact, when discussing the first proposals of the GDPR, the A29WP noted that "the risk-based approach is being increasingly and wrongly presented as an alternative to well-established data protection rights and principles, rather than as a scalable and proportionate approach to compliance".[27]

The HLEG on AI (see below) noted that "AI is not an end in itself, but rather a promising means to increase human flourishing".[28] While other ethical frameworks can be adapted to AI regulation, the proposal made by the HLEG seems inspired by Kantian spirit, deontological ethics and the promotion of human dignity.[29] Consolidated principle-based approaches in bioethics have served as a conceptual basis for the development of AI ethics.[30] A certain degree of convergence around the four bioethical principles of Benefice, Non-Maleficence, Autonomy and Justice can be found in the conceptual framework adopted by AI4People, which first found agreement among previous sources (including the Asilomar AI Principles and IEEE initiative).[31] In accordance with this position, bioethics is consistent with the onto-centric, patient-oriented, ecological premises of information ethics.[32] In turn, it has been endorsed by the HLEG and consequently taken into account by the EU Commission's White Paper. The OECD has recently shared the bioethical perspective

25 Sachs (2016)

26 Veale and Borgesius (2021)

27 Article 29 Working Party, Statement on the role of a risk-based approach in data protection legal frameworks (30 May 2014) https://ec.europa.eu/justice/article-29/documentation/opinion-recommendation/files/2014/wp218_en.pdf

28 HLEG (2019)

29 Petit and De Cooman (2021)

30 See Mittelstadt (2019)

31 See the review by Beauchamp et al. (2001)

32 See Floridi (2013), Chap. 4.5

in the document "Recommendation of the Council on Artificial Intelligence" (not included among our sources)

Mittelstadt[33] has welcomed the adoption of medical ethics principles by claiming its convenience "as it is historically the most prominent and well-studied approach to applied ethics". Advantages consist of an inclusive discourse that keeps into account the needs of all the actors involved, some degree of flexibility when contextual-dependent balances are needed, and the capability of framing ethical challenges and provide clinical decision-making guidance. AI ethics seems intended to add some kind of normative content addressed to decision-makers and AI designers. However, other commenters underlined the inadequacy of bioethical approach, as "principles of bioethics can impose excessive burdens on AI systems if applied without paying heed to the principle of proportionality or giving guidance on how principles will be endorsed or enforced".[34]

Attempts have been by the "AI Charters" discussed below, which have adopted similar approaches by listing principles that should be embraced when designing, developing and deploying AI systems. However, the AI Act proposal seems to opt for a different regulatory strategy when adopting the risk-based compliance model described above. Nonetheless, traces of a principle-based approach can be found in Recital 15[35] and Recital 17.[36]

Recital 28 of the AI Act proposal seems to clarify the relationship between risk and fundamental rights.[37] The adverse impact on AI technologies is measured by taking into account how compromised fundamental rights would be by their deployment.

As mentioned in the introduction to this subsection, our Blueprint embeds both approaches. So far, this book has devoted efforts to describe the risks that can emerge from an unregulated adoption of machine learning systems in food safety. While departing from a risk-minimisation approach, the Blueprint shall also

---

[33] Mittelstadt (2019)

[34] Renda et al. (2019), p. 47

[35] Aside from the many beneficial uses of artificial intelligence, that technology can also be misused and provide novel and powerful tools for manipulative, exploitative and social control practices. Such practices are particularly harmful and should be prohibited because they contradict Union values of respect for human dignity, freedom, equality, democracy and the rule of law and Union fundamental rights, including the right to non-discrimination, data protection and privacy and the rights of the child

[36] AI systems providing social scoring of natural persons for general purpose by public authorities or on their behalf may lead to discriminatory outcomes and the exclusion of certain groups. They may violate the right to dignity and non-discrimination and the values of equality and justice [...]

[37] "[...] The extent of the adverse impact caused by the AI system on the fundamental rights protected by the Charter is of particular relevance when classifying an AI system as high-risk. Those rights include the right to human dignity, respect for private and family life, protection of personal data, freedom of expression and information, freedom of assembly and of association, and non-discrimination, consumer protection, workers' rights, rights of persons with disabilities, right to an effective remedy and to a fair trial, right of defence and the presumption of innocence, right to good administration [...]"

be focused on ensuring that adequate relevance is given to principles capable of minimising these risks. To put it differently, the compatibility of the two approaches can be realised by stressing that adherence to certain principles can effectively minimise the risks.

Adherence does not equal to "ethics only" regulatory model.[38] Instead, implementations would be required to ensure that these principles are translated into practice. Unfortunately, there are no established mechanisms to perform such translation[39] and only possible ideas for implementations can be proposed.[40] These proposals may only mitigate the main limitation of adherence to certain principles, i.e., their non-binding nature.

5.1.3 Institutional AI Charters

Let us introduce the primary sources of our investigation on AI principles. They consist of AI ethical Charters published by EU and Member States institutions. Among the latter, Germany, France, the Netherlands, and Italy have been selected for their institutional activities on AI governance. The United Kingdom has been excluded due to its recent withdrawal from the European Union.

5.1.3.1 European Commission Communications: "Artificial Intelligence for Europe" and "White Paper On Artificial Intelligence"

European Commission's Communication "Artificial Intelligence for Europe"[41] lays down the EU approach to an effective and trustworthy deployment of AI technologies in Europe. The document covers a wide range of topics, including the way in which AI is already changing our life, the research directions, and key areas of investments. This document comes at the end of an EU-wide dialogue that involved a cooperation agreementand the setting up of a High-Level Expert Group (HLEG, discussed below) in 2018, whose conclusions were endorsed in a subsequent Communication by the Commission in 2019.[42] Moreover, these

[38] Petit and De Cooman (2021)

[39] Mittelstadt (2019)

[40] This will be the task of Sect. 5.3

[41] Communication From the Commission to the European Parliament, the European Council, the Council, the European Economic and Social Committee and the Committee of the Regions "Artificial Intelligence for Europe" (OM/2018/237 final)

[42] EU Commission, "EU Member States sign up to cooperate on Artificial Intelligence"; Communication from the Commission to the European Parliament, the Council, the European Economic and Social Committee and the Committee of the Regions Building Trust in Human-Centric Artificial Intelligence (COM/2019/168 final)

documents have to be conceived as preliminary works for the 2020 Commission White Paper on Artificial Intelligence.[43,44]

Taken together, these policy documents highlight a clear direction for EU AI[45] and data governance, whose aim is minimising risks, promoting opportunities and creating a trustworthy ecosystem. In its development, the EU approach to AI seems to progressively endorse a regulatory trend, yet limited for AI systems that pose high risks (White Paper, p. 3). Threats for safety, consumer and fundamental rights might derive from the sector in which AI is deployed (e.g., healthcare, transport, energy, etc.) and the characteristics of the system itself (e.g., the possibility to produce legal effects on individuals). These two criteria identify sensitive areas that call for detailed forms of regulation (White Paper, p.17). As regards areas other than the critical ones, the Commission has progressively refined its approach without explicitly calling for "hard" regulation. For these areas, we can rely on the observation by Pagallo and colleagues,[46] according to whom "the debate is about how to complement and strengthen the existing regulation".

From a technical perspective, machine learning is defined as a subset of AI in which "algorithms are trained to infer certain patterns based on a set of data in order to determine the actions needed to achieve a given goal" (White Paper, p. 16), hence the necessity of adapting such high-level AI framework to the peculiarities of the technicalities that we described in Chap. 2 of this book. *Mutatis mutandis* some technical considerations of the Commission shall be deemed extendible to the technical advancements observed in our domain.

From a legal perspective, the food safety domain is not an area that meets the two criteria used to identify areas that pose risks to individuals.[47] We already noted that EFSA's opinions neither have a significant impact on individuals, nor do they directly pose risks to fundamental rights. Instead, according to the way EFSA's conclusions are substantially and procedurally drawn, they tend to take into account groups of individuals that present similar eating behaviours and are inserted within decision-making processes for which EFSA *supports* decisions that might impact on fundamental rights (in particular, the right to health). More importantly, the food and chemicals industries are never discussed in terms of "AI high-risk" market sectors,

[43] EU Commission, "White Paper On Artificial Intelligence—A European approach to excellence and trust" (COM/2020/65 final)

[44] See Pavon and Gonzalez-Espejo (2020), Chap. 2.02A

[45] Artificial Intelligence is described as "systems that display intelligent behaviour by analysing their environment and taking actions—with some degree of autonomy—to achieve specific goals" (Artificial Intelligence for Europe, p. 1)

[46] Pagallo et al. (2019)

[47] Instead of excluding *a priori* any interest in agri-food safety due to the lack of (high) risks, identifying criteria for a trustworthy deployment of machine learning techniques is still necessary due to the absence of trust in the system already shown and discussed. What is at stake is not the necessity of regulating the deployment of machine learning, but how a principle-based discussion can foster trust in the system

thus preventing any expectation of tailored regulatory solutions for the industrial side of our domain.[48]

The EU Commission has reached an agreement on seven core principles to steer future regulation of high-risk AI systems.

- Human agency and oversight. The Communication calls for control measures over the functioning of AI systems. Adaptability, accuracy and explainability of AI machines are mentioned together with human-in-the-loop, human-on-the-loop, or human-in-command models.[49]
- Technical robustness and safety. Security, reliability and robustness are key to ensure the capability of the systems in the handling of errors or fallacies. Two necessary prerequisites—reproducibility of the results and safety/security-by-design approaches—are also emphasised;
- Privacy and data governance. Data protection and privacy are associated with the need of constructing databases that do not reflect biases or generate inaccuracies. As noted by commentators, this approach calls for a new understanding of data protection tools that is detached from the individual and transactional conceptualisation of the current regulatory framework and can synergise with other principles, such as human oversight, transparency, and fairness;[50]
- Transparency. Algorithmic transparency comes with the traceability of AI systems (log, documents, data gathering and labelling procedures) and explainability/explicability of the logic underlying the decision, as well as other factors that led the deployment of the algorithms (e.g., business model);
- Diversity, non-discrimination and fairness. Datasets can be imbalanced or poor quality and produce biases and incompleteness. Design, dialogue and participatory solutions are necessary to include all the stakeholders that might be affected by the system;
- Societal and environmental well-being. This principle relates to the societal and environmental externalities of AI systems, to be understood also in light of UN SDGs;
- Accountability. This principle should ensure auditing (*ex-ante*) and redress (*ex-post*) mechanisms to identify potential risks and to prevent lack of responsibility for the use of AI systems.

These principles have merged into the proposed AI Act discussed above (Sect. 5.1.2) in various degrees and according to the risk classification. Societal and environmental well-being are traversal to all AI applications (Recital 3), whereas privacy is a general concern (Recital 15). Human oversight is a specific measure for high-risk AI systems (art. 14), like technical robustness (art. 15) whereas

[48] This is an additional justification for the "soft ethics" nature of the Blueprint presented in this thesis

[49] For further insights on this model, see also Quintarelli et al. (2019) and the description made by the HLEG

[50] Kuner et al. (2018)

transparency is expanded to limited-risk AIs (art. 52). Non-discrimination is a general concern (Recital 15), mitigated by a specific and newly introduced legal basis for the processing of personal information, i.e., debiasing (art. 10(5)).

5.1.3.2 German Federal Government: "AI Strategy"

The German approach to AI is described by a joint documents signed in 2018 by the Ministries of Education & Research, Economic Affairs and Energy, Labour and Social Affairs.[51] The Strategy defines AI from the well-known "strong" vs "weak" perspectives.[52] The document does not present an explicit list of values that corroborate the existing legislation and calls for amendments of the current regulatory framework when modifications seem needed following the necessary evaluations (p. 39). However, the document shows adherence to ethical principles in consistency with German "liberal democratic constitutional system" (p.8). Such integration should be realised by a broad societal dialogue. It is then possible to identify some steering principles throughout the Charter:

- (Data) Security. It is presented as a functional component of the societal benefit, both for the collectivity and for German attractiveness for export (p.8)
- Explainability. This principle is understood as a way to prevent opacity of "black-box" algorithms. Explainability is functional to ensure trust in AI systems and to assess their compliance with legal requirements (p. 16)
- Transparency. This principle is partly overlapping with explainability as regards its goals (promoting trust by accessing the functioning of the model, p.16), but it seems mostly referred to the results of the whole algorithmic process (p.10) rather than the features of the underlying mathematical or logical model. Specifically, it is conceived as a way to prevent and challenge undue discrimination, both at individual and group level (p.38). When considering this goal, other dimensions of transparency—predictability, non-discriminatory nature and verifiability—are recommended in the "development, coding, introduction and use of AI systems (including training and application data)."
- Accountability. Trustworthiness is also fostered by the principle of accountability, which calls for measures introducing auditing, developing impact assessment standards and the disclosure of AI algorithms (p.38).
- Privacy and Data Protection. The German AI Strategy is keen on promoting "self-determination (particularly the right to control one's data)" and citizens' privacy (p.39). In the area of privacy and data protection, careful recommendations are

---

[51] German Federal Government, "Artificial Intelligence Strategy" available at https://www.ki-strategie-deutschland.de/home.html?file=files/downloads/Nationale_KI-Strategie_engl.pdf, accessed on 27/04/2022, hereafter BMWi (2018). Please note that the strategy has received an update in 2020. However, no significant changes for the purposes of this study can be found

[52] Russell and Norvig (2010), Chap. 1. The Strategy also explicitly mentions machine learning approaches within its scope (p. 5)

made, including the development of synthetic training data to avoid bulk data collections.

5.1.3.3 French Conseil National du Numerique: "For a Meaningful Artificial Intelligence: Towards a French and European Strategy"

The French AI strategy document—"For a meaningful artificial intelligence: Towards a French and European strategy",[53] also known as Villani Report from its main contributor—points out a significant number of ethical principles to steer the adoption of AI systems. Machine learning is explicitly covered by the charter as it is deemed one of the key advancements of AI (p.4). Part 5—"What are the Ethics of AI?"—illustrates a significant number of ethical principles to follow in the development and deployment stages of AI systems. Principles are needed to maximise the collective benefit of French and European societies. Ethical recommendations are intended as principles that "occupy the available space between what has been made possible by AI and what is permitted by law, in order to discuss what is appropriate". However, also the embodiment of ethically-driven solutions in the design ("ethics-by-design") is a recommended method of practical implementation.

It is possible to extract five core principles:

- Transparency and Auditability. Explicit references are made to the "black box" problem and the need of ensuring explainable AI decisions (pp. 114–115), also by means of promoting research (p.118). Explainability is directly linked to the accountability of decisions that have an impact on fundamental rights and freedoms. The document straightforwardly affirms "as a society, we cannot allow certain important decisions to be taken without explanation" (p. 115). Transparency and auditing solutions are then proposed. Regarding the former, it is noted that "businesses that have invested substantial sums of money in the construction of their algorithmic systems and would like to reap their rewards are necessarily reluctant to see their intellectual property divulged to third parties"(p.117); then, a solution in public auditing done by expert commissions is proposed alongside incentives to ensure a wide access to data and algorithms also for the purposes of validating legal and ethical compliance of the AI systems;
- Fairness. Tackling discrimination seems a major concern (p.113), as "[t]he use of deep learning algorithms, which feed off data for the purposes of personalization and assistance with decision-making, has given rise to the fear that social inequalities are being embedded in decision algorithms" (p.116). For these purposes, an *ex-ante* "Discrimination Impact Assessment" is proposed as a self-evaluation intended to prevent unfair outcomes of AI systems deployment.

[53] Hereafter, Villani et al. 2018

- Accountability and Explainability (as research areas). The responsibility correlated to the use of machine learning systems is defined as a "a real scientific challenge" due to the necessary balance that has to be found between the necessity of explaining automated decisions and the efficiency of algorithms used to make them. In fact, while some algorithms are more explainable—yet, less precise—than others (p. 115), the social acceptance of certain automated decisions might be low if no justification is provided. Therefore, research into accountability and explainability is recommended (p. 118);
- Privacy and (group) data protection. One of the most innovative and perhaps controversial fields of the Villani Report is the provision of collective rights to data. First, the importance of aggregated data for AI is juxtaposed to the "blind spots" of EU and French data protection law. As aggregated data fall outside the scope of data protection law (as far data are no longer referred to identified or identifiable individuals), the collective dimension of the effects of decisions on groups is left unprotected (p. 121). A possible solution is found in the implementation of collective or class actions supported by compensation for injuries (p.122);
- Inclusiveness. Given the importance of the ethical questions at stake, an inclusive debate is recommended. Following the German Data Ethics Commission, a national advisory committee is proposed as a *forum* for coordinating such ethical discussion (p. 128) by mediating the positions of the industry, institutions, NGOs, academia, trade unions, etc. Institutional and public consultations via surveys and opinion polls would fall within the competences of this committee (p. 129).

5.1.3.4 SIGAI, The Special Interest Group of AI: "Dutch AI Manifesto"

The Dutch Artificial Intelligence Manifesto[54] originates from academic researchers operating in the area of AI. The technical sections of the document reflects its nature by discussing cutting-edge AI applications, including machine learning. Despite not presenting an explicit list of principles, some morally-relevant *desiderata* can be retrieved from some passages.

- Fairness. This goal is briefly mentioned when discussing the trade-off between algorithmic efficiency of reinforcement learning algorithms (p. 12). Fairness seems related to the social awareness of AI, i.e., its capability of integrating and interacting with human-populated environments. As a consequence, machine learning systems should not provoke negative effects on individuals when analysing human behaviour, group interaction patterns, and emotions. This dimension is included in the Socially-Aware AI paradigm, i.e., the promotion of a collaborative AI that is capable of interacting with humans, interpreting and

[54] Hereafter, SIGAI (2019)

positively influencing human behaviour while coordinating its interaction with individuals.

- Explainability. As with other documents, Explainability is presented as a challenge that regards the opacity of "black box" algorithms and the effects of automated decisions on individuals (p. 14). On the one hand, research is needed to ensure that available machine learning techniques guarantee a sufficient understanding of their internal structure; on the other hand, users shall be able to interact with the system to grasp the "why and how" the decision has been taken by means of intelligible user interfaces.
- Accountability. Presented within the context of "Responsible AI", accountability refers to the *ex post* validation of the behaviour displayed by an AI system from a moral and legal perspective. Responsible AI mitigates the risks emerging from a wide range of scenarios—from algorithmic biases to privacy-invading technologies—that require the adoption of an ethical aptitude and the respect of normative standards. Accountability is then needed to trust the use of these systems and their convergence towards value-oriented behaviours.

One commentator[55] has underlined the multidisciplinary character of the Dutch AI Manifesto when it comes to the three *desiderata* mentioned above. With this regard, the relationship between social, explainable, responsible AI and the efforts of the legal informatics academic community in promoting AI & Law conferences such as ICAIL and JURIX have been underlined. Other reviewers[56] have underlined the role of Sustainability, a key dimension of environmental social awareness.

5.1.3.5 AGID, Agency for Digital Italy: "White Paper on Artificial Intelligence at the Service of the Citizen"

The Italian AI strategy has been preliminary drafted by the "White Paper on Artificial Intelligence at the service of the citizen" released by a task force within the National Agency for Digital Italy.[57] The document mainly pertains to the adoption of AI systems by public institutions and its inclusion seems convenient in the light of our analysis. Ethics and principles are identified among the "Challenges at the service of citizens" and are displayed on top of the list. Principles endorsed by the document represent a convergent positions between the AI "enthusiasts", i.e., those who propose a straightforward adoption of AI systems at all levels with no conditions, and "doom-mongers", i.e., those who sceptically reject the integration of AI systems within public administrations due to possible threats to individuals' rights and freedoms.

---

[55] Verheij (2020)

[56] Ryan and Stahl (2020)

[57] Hereafter, AGID (2018)

Four elements of debate and *desiderata* are presented as a way to commit to an adoption of AI systems for the public sector that is beneficial for the general public.

- Data quality and neutrality. This principle correlates to a fair construction of datasets, in particular to avoid the introduction or the replication of errors and biases that might open the room for discrimination (ethnic differences in crime prevention AI systems are mentioned). The over/underestimation of the weight of certain variables is likely to raise issues regarding the interpretation of machine-generated previsions. Non-discrimination is also discussed in Challenge 7 "Preventing Inequalities", which suggests a proactive approach for which AI systems should minimise the existing differences rather than allowing their increase.
- Responsibility (accountability and liability). This principle covers both the legal liability for the use of AI systems by public bodies, but also the moral dimension of accountability. Some of the issues discussed "highlight the need to establish principles for the use of AI technologies in a public context" and the political responsibility of AI-supported decisions.
- Transparency (and Openness).[58] These principles are "fundamental prerequisite" for a trustworthy adoption of AI systems by public decision-makers in light of the objective of avoiding discrimination and information asymmetries. Therefore, transparency is needed to guarantee citizens' right to understand public decisions. It is remarked that algorithms could introduce social discrimination. However, explicit recommendations (e.g., explainability, auditing, disclosures) are missing and it is fairly hard to grasp the essential meaning of transparency for the purposes of this document.
- Protection of the private sphere. While this principle seems broader that data protection as it might refer to non-informational dimensions of privacy, the discussion that follows is mainly focused on the protection of personal data. In particular, public bodies should commit themselves to protect citizens' data, in particular those sensitive, and when data are used in contexts other than the original one for which they have been collected.

Due to its limited scope, namely the deployment of AI in public administration, the advantage and the main drawback of the Italian AI strategy consists of its narrow remit. On the one hand, it contributes to enrich the institutional perspective by analysing *desiderata* of public bodies. On the other, it lacks the generalisability/universality that marks other Charters.

On July 2020, the Italian AI strategy was enriched and finalised by an other document[59] listing several policy recommendation. While maintaining the human-centric premises of the previous version, this new document highlights the necessity

---

[58] As no further references of "openness" are provided in the document, I would cautiously assume that the term is used *ad abundantiam* to support transparency. As such, the two terms are used as synonyms

of collaborating to the debate on AI Trustworthiness (p. 24) and recommends the Government to carefully align with to the EU AI ethics vision and its human-centric approach (p. 87).

5.1.4 Interdisciplinary Ethical Charters

Auxiliary sources consist of documents that have been drafted and released in non-institutional contexts, namely a multi-stakeholder forum (Asilomar Conference), an engineering and informatics non-profit organisation (the Institute of Electrical and Electronics Engineers (IEEE)), an academic-driven group (AI4People) and the High Level Expert Group (HLEG) on AI set out by the EU Commission. These document will be used as additional resources to identify the origins of the aforementioned institutional charters and to approach the current debate on AI governance from a broader perspective.

5.1.4.1 Asilomar Conference

The Future of Life Institute 2017 Beneficial AI Conference, also known as Asilomar Conference,[60] is one of the first modern examples of international and multi-stakeholder[61] discussion on AI ethics. Following a 2.5 day debate, the Conference produced a list of principles as its outcome. The list has been drafted starting from existing reports released by academic, governmental, industrial and non-profit entities. Simplifications, summaries and surveys, both at individual and group level, were then carried out. Finally, the list of 23 principles on which a 90% agreement had been reached was adopted.

Focusing on principles 6–18 ("Ethics and Values"), it is possible to cluster them into some groups:

- Beneficence. Principle 14 ("Shared Benefit") and 15 ("Shared Prosperity") promote the collective benefit of AI development;
- Accountability. Principle 9 ("Responsibility") calls for moral thinking to be requested to AI developers and designers regarding the use and misuse of AI technologies. Then, Principle 10 ("Value Alignment") sets the direction for moral thinking by requiring that AI is designed to operate according to human values;
- Transparency. Principle 7 ("Failure Transparency") and 8 ("Judicial Transparency") specify that it is necessary to design systems that are transparent

[59] To date (April 2022), an English translation of the document is not yet available. The original version is available at https://www.mise.gov.it/images/stories/documenti/Proposte_per_una_Strategia_italiana_AI.pdf, accessed on 27/04/2022

[60] Future of Life Institute (2017)

[61] To date (April 2022), Asilomar Principles have been signed by 1797 AI/Robotics researchers and 3923 other signatories

when failing and provide explanations when operating. However, the second transparency measure is limited to AI systems deployed in the context of judicial decision-making;

- Security. Principle 6 ("Safety") requires that AI systems are safe and secure when operating. No further specifications are offered as regards the nature of such safety (e.g., data, physical infrastructure, non-discrimination). Principle 7 ("Failure Transparency") completes the principle of Safety with an *ex-post* perspective on the necessity of preserving the possibility of scrutinising the reasons for mistakes;
- Control. Taken together, principles 10 ("Value Alignment"), 16 ("Human Control"), 17 ("Non-subversion"), 18 ("AI Arms Race") strongly encourage the development of AI systems that can ensure human oversight from different perspectives, all united by ensuring the promotion of human values. Firstly, humans should be free to rely or not on AI systems for decision-making processes; secondly, AI should not threaten human health; thirdly, AI arms race should be avoided;
- Privacy. Principles 12 ("Personal Privacy") and 13 ("Liberty and Privacy") reflect the threats to personal data and individual rights and freedoms that rise when discussing the implications of AI systems. While the former principle is mostly focused on giving the right to "access, manage and control" personal data, the latter takes the broader perspective of protecting individuals' "real or perceived liberty".

5.1.4.2 IEEE Ethically Aligned Design

Within the context of IEEE Global Initiative on Ethics of Autonomous and Intelligent Systems, IEEE has published two versions of "Ethically Aligned Design" in 2016 and 2017,[62] aiming at establishing "ethical and social implementations for intelligent and autonomous systems and technologies, aligning them to defined values and ethical principles that prioritize human well-being in a given cultural context." (p. 2). The document purposively covers a wide range of AI technologies and analyses their ethical implications from the viewpoint of researchers, manufacturers, and designers of intelligent and autonomous systems, with specific focuses on lethal autonomous weapon systems and "strong" AI applications.

An explicit list of five principles is then identified:

- Human Rights. AI should respect internationally recognised human rights enshrined by the Declaration of Human Rights, UN treaties and international conventions. Fundamental rights and freedoms shall be preserved in any AI

[62] See IEEE (2017). In our analysis, we will refer to the Version 2 of the document, available at https://standards.ieee.org/content/dam/ieee-standards/standards/web/documents/other/ead_v2.pdf, accessed on 27/04/2022

application. This entails the preservation of safety in the runtime of the system and the traceability of possible failures;

- Well-being. Crucially, IEEE points out that there are no established metrics of well-being, hence the possible failure of attempts made to promote it in the development of AI applications. Therefore, the promotion of well-being has to be contextualised to certain metrics, even beyond Gross Domestic Product (GDP);
- Accountability. As with other charters, IEEE also focuses on the necessity of holding some entities responsible for the harms caused by intelligent and autonomous systems. Since no specific entity can be found *a priori* liable, a distributed form of apportion culpability among manufacturers (which carry the burden to prove the correct functioning of the system), designers, operators, owners. This is deemed necessary to ensure a trustworthy and widespread adoption of AI systems;
- Transparency. This principle ensures that "it is possible to discover how and why a system made a particular decision, or in the case of a robot, acted the way it did". The principle of Transparency has three nuances: traceability, explainability, and interpretability. These facets are needed to understand what the system is doing, validate and certificate AI systems, investigate malfunctioning, carry out legal investigations, increase trust towards AI systems (in particular, disruptive technologies);
- Awareness of Misuse. AI designers and programmers, citizens, and decision-makers should be educated towards a responsible use of AI systems focused on the development of accountability mechanisms.

5.1.4.3 AI4People Initiative

The promotion of a "Good AI Society" is the foundational scope of an other international initiative, AI4People within the Atomium-EISMD consortium.[63] The article first discusses opportunities and risks of AI with a focus on human dignity; then, it identifies five principles that should steer AI development and deployment; finally, it elaborates 20 principle-based recommendations that could promote a responsible adoption of AI systems.

Methodologically speaking, the AI4People initiative identifies ethical principles taking inspiration from bioethics[64] and previous works done in the context of AI ethics.

- Beneficence. Taking inspiration from previous works in AI ethics, AI4People authors synthesise existing principles as the need of creating AI technologies are beneficial for humanities and promote the "well-being of people and the planet";

[63] Floridi et al. (2018)

[64] As the authors state: "Of all areas of applied ethics, bioethics is the one that most closely resembles digital ethics in dealing ecologically with new forms of agents, patients, and environments". The conclusion was previously reached in Floridi (2013)

- Non-maleficence. This principle calls for the avoidance of negative consequences linked to the misuse of AI systems. Risks might emerge in the field of privacy and data protection, arms race (as already pointed out by the Asilomar Principles), or in other sectors discussed in other ethical documents. In AI4People, the principle of non-maleficence as harm prevention encompasses both accidental and deliberate harms, respectively referred to as "overuse" and "misuse". Moreover, both human (e.g., developers) and machine behaviours are subject to this principle;
- Autonomy. AI4People authors found large consensus on ensuring the freedom to choose to rely on human or automated decision-making processes. Interestingly, they propose a "static" perspective, which promotes the prominence of human autonomy, and a "dynamic" perspective, according to which the autonomy of machines should be made "intrinsically reversible" and human subversion shall be guaranteed following an initial delegation to AI. Individuals should also be free to choose whether or not to rely on AI systems ("meta-autonomy") for reasons overriding efficacy;
- Justice. In works before AI4People, the social implications of AI are discussed under several dimensions, including a proactive aptitude to reduce discriminations, sharing benefits of the adoption of AI technologies, preventing unknown harms. By acknowledging these dimensions consistently with their bioethical approach, AI4People authors identifies "justice" as the need of ensuring that AI-related resources and benefits are distributed widely;
- Explicability. Differently from other principles, explicability is not derived from bioethics. It specifically relates to AI and constitutes one of the most prominent elements of novelty in AI4People's research. Previous works used different terms to express the ability to understand AI systems inner logic ("transparency") and to identify the responsible entity ("accountability") for the use of these technologies. In addition to them, AI4People authors conceptualise Explicability both in the epistemological sense ("intelligibility", or the making sense of the operations performed by AI systems) and in the ethical one ("accountability", or the identification of a responsible entity for the work done by the AI).

5.1.4.4 EU Commission High Level Expert Group: "Guidelines for Trustworthy AI"

The EU Commission High Level Expert Group (HLEG) is one of the most discussed institutional initiatives on AI Ethics.[65] Its Guidelines for Trustworthy AI set out a framework for achieving a lawful, ethical and robust AI systems' life-cycle. Therefore, as the authors state, their outcome aims to go "beyond a list of ethical principles" (p.2). Such framework consists of three essential components:

[65] See HLEG (2019). The Expert Group was set up by the EU Commission in June 2018. As stated in the foreword, the content of the document does not reflect the official position of the Commission, hence the inclusion among our auxiliary sources

four ethical principles, seven key requirements, and an assessment list to verify adherence to principles and requirements (Fig. 5.3).

Ethical principles (in yellow), key requirements (in green) and their details (in orange) are displayed in the picture below. Ethical principles are based on fundamental rights enshrined in the EU Treaties, the Charter and international right law (p. 9). The consist of:

- Respect for human autonomy. This principle mirrors human dignity and the right to self-determination and prescribes that, when using AI systems, humans shall not be manipulated or deceived by the system. When developing AI technologies, such human-centric approach shall be ensured by securing human oversight, i.e., the ability to control the system by various degrees;[66]
- Prevention of harm. AI may cause or exacerbate harms to human dignity or physical or bodily integrity. Therefore, it is necessary to mitigate threats to human beings by ensuring that AI systems are secure and resilient. This principle consists of several dimensions which range from the protection of data integrity (including personal data), to their quality and accuracy. In particular, the HLEG recommends keeping into account vulnerable people, situations of informational asymmetries and natural environment when identifying risks to mitigate;
- Fairness. This principle has a substantive and a procedural dimension. From the substantial perspective, fairness implies (a) that unfair and discriminatory AI-supported decisions shall be avoided, and (b) that proportionality between means and ends in the use of AI systems is kept into account by AI practitioners. From a procedural perspective, Fairness shall grant the possibility to seek redress for the automated decisions made by AI algorithms, accountability mechanisms for the use of AI systems and explicability of AI decisions;
- Explicability. Juxtaposed with "black box" algorithms, this principle pertains to the design of AI systems capable of communicating their processes and their output, also in relation to the context of deployment and the risks generated by the outcomes. Explicability is needed for a plethora of reasons: to promote trust in AI systems, to contest AI-supported decisions, to verify compliance with fundamental rights.

The work done by the HLEG has been criticised by one of its members—Thomas Metzinger—because of the group's composition, allegedly too unbalanced towards non-ethicists (48 vs 4 ethicists) and its proximity to the industry which was reflected in censorship.[67] These claims are undoubtedly out the scope of this dissertation

---

[66] The HLEG discusses three alternatives (p. 16): human-in-the-loop (HITL), human-on-the-loop (HOTL), and human-in-command (HIC). HITL refers to the capability of controlling every operation of the system for a human; HOTL is the ability to control the design of the operations (*ex ante*) and monitor them (*ex post*). HIC is the ability to scrutinise and possibly steer the large-scale effects of the systems, including societal ones

[67] Metzinger's claims were made by means of an article at Der Tagesspiegel available at https://www.tagesspiegel.de/politik/eu-guidelines-ethics-washing-made-in-europe/24195496.html, accesses on 27/04/2022

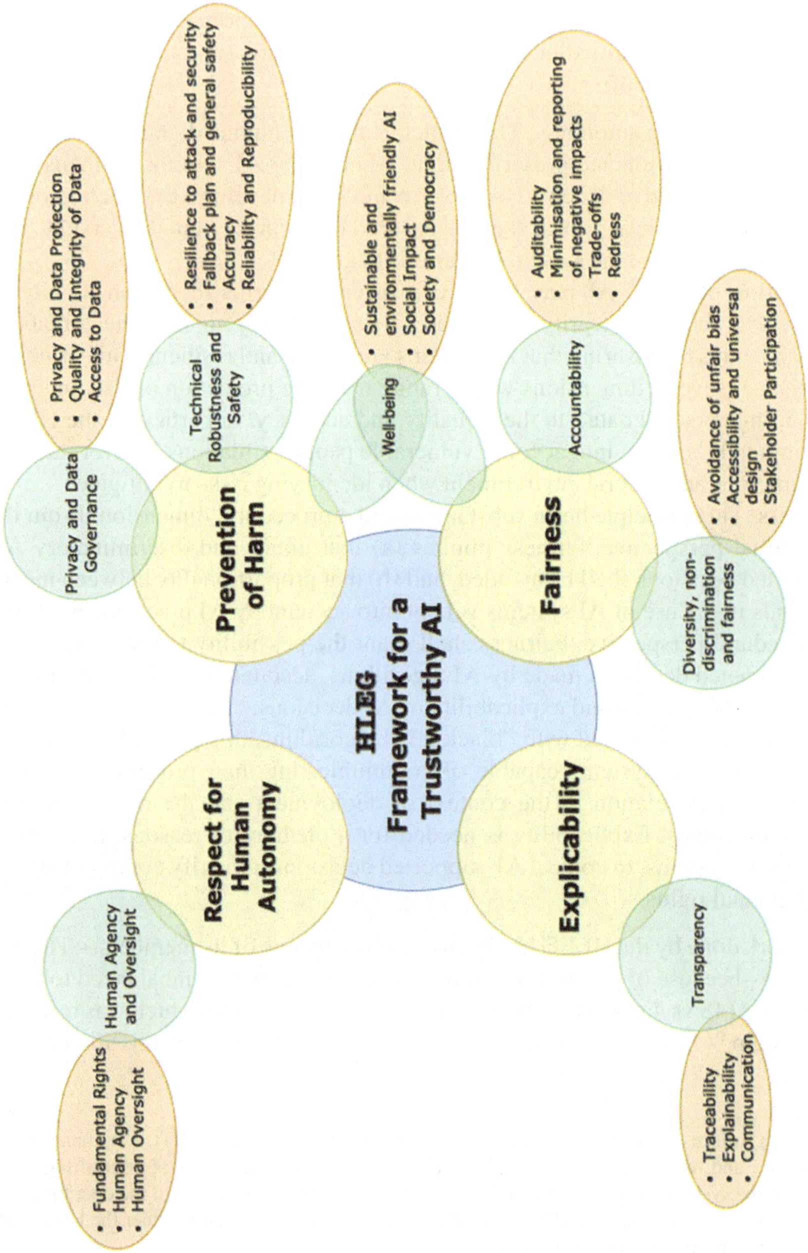

Fig. 5.3 EU Commission HLEG on AI—key findings

and we will leave them to scholars more interested in gossip than advancements of academic debate.[68] As regards their contents, the Guidelines have been defined as "lukewarm, short-sighted and deliberately vague" and, at the same time, "the best globally available platform for the next phase of discussion" by Metzinger. Other scholars have criticised the absence of certain rights and freedoms (e.g., the freedom to assembly) and the appearance of topics placed outside EU Charters and Treaties, e.g., sustainability.[69]

5.2 The P-SAFETY Blueprint for Trustworthy Food Safety

The Blueprint can be defined as an introductory ethical framework to integrate, interpret and align the food safety risk assessment regulatory system applicable to data to the principles guiding the deployment of AI (*rectius*, machine learning techniques) that have emerged from the policy documents and the academic discussion reported above. On their bases, the Blueprint proposes as a set of general principle-based recommendations to support decision-making processes related to data, algorithms and practices of the domain at stake.

It is possible highlight some features of the Blueprint departing from this concise summary. The Blueprint:

- is an *introductory* framework to be further analysed and implemented. Although some recommendations will be provided throughout the Blueprint and in the next Section, a detailed assessment on practical implementations will be left to future research;
- is an *ethical* framework since it identifies values and principles that derive from contributions in AI and data ethics;
- can be used to *integrate, interpret and align* the existing regulatory framework in light AI principles, rather than replacing it. Once the proposed AI Act will be approved, the Blueprint will serve as a trait d'union between food law and the data-related provisions of such regulatory framework;
- fosters integration, interpretation and alignment of *data-related* provisions contained in the EU food safety risk assessment regulatory framework. Therefore, its scope is narrower than the one of food safety regulation and mainly concerns the processing of data for risk assessment purposes;

[68] However, it is worth reminding that the decision-making process of the HLEG was open to public feedbacks, available at https://digital-strategy.ec.europa.eu/en/news/over-500-comments-received-draft-ethical-guidelines-trustworthy-artificial-intelligence, accessed on 27/04/2022: more than 500 comments were received over the first draft and significant changes were made before the final publication

[69] Wagner (2018)

- fosters integration, interpretation and alignment of (mainly) *machine learning techniques* insofar they are deemed to fall within general notion of AI for regulatory purposes;
- is grounded on *policy documents and academic literature*. The Blueprint is based on technical, legal and ethical contributions, on the one hand, and EU and Member States AI charters, on the other;
- adopts a *principle-based* approach. Noteworthily, principles are indirectly related to bioethics;
- proposes recommendations to *support decision-making processes* at every level, including institutional policy-making or industrial self-regulatory initiatives;
- supports decision-making processes related to *data, algorithms and practices* consistently with the data-centric approach selected as the main technical standpoint.

The proposed list of principles is *non-exhaustive* and *non-hierarchical*. Therefore, other elements might be added in future research, either in light of advancements in legislation or for practical purposes. Moreover, the list shall not be interpreted as a ranking[70] and therefore it is crucial to assume principles enshrined in the Blueprint might conflict either *internally* among them or *externally* with other principles and values, especially in concrete cases and when no *lex superior derogat inferiori* rule could be used.[71] When this is the case, one possible unified approach to solve such conflicts might be the Laws of Balancing presented by Alexy,[72] also used in the context of legal argumentation.[73] The twofold internal and external relevance of Alexy's Laws of Balancing seems appropriate for dealing with conflicting scenarios.

[70] The nature of privacy as a meta-principle will be clarified in the appropriate sub-section

[71] As in the attribution of weights to the principles in a manner that allows the prevalence of one on others discussed in Alexy (2000)

[72] Alexy (2003). The first Law of Balancing, also known as the Substantive Law, states that "[t]he greater the degree of non-satisfaction of, or detriment to, one right or principle, the greater must be the importance of satisfying the other" (substantive importance of reasons). The second Law of Balancing, also known as the "Epistemic Law", states that: "[t]he more heavily an interference with a constitutional right weighs, the greater must be the certainty of its underlying premises". See also the discussion in Sartor (2010) on the role of goal-norms in legislator's teleological reasoning and benchmarks for balancing and proportionality. For an applied reasoning in AI & Law see Lauritsen (2015)

[73] Feteris (2017), Chap. 7.5

5.2.1 Security

Security encompasses several dimensions, all related to issues regarding external threats to information systems.[74] Threats can occur from the interaction of unauthorised third parties with the information system at stake.[75]

In the domain of interest, security pertains to a wide range of areas, including the technical protection of data warehouses, confidential private data, food consumption and background information (data security). At the same time, it also regards the technical robustness of AI systems and their capability to avoid fallacies, to be implemented in light of the *by-design* approach (AI security).

As regards data security, information sharing is desirable for the cross-validation of scientific data and often necessary for legal compliance. If, as argued, information exchange is also based on trust, a legitimate expectation of security is likely to arise among the involved entities. For these purposes, the functional role of security in ensuring trust noted by one institutional charter has to be remarked.[76]

In food safety risk assessment, data security follows the well-know dimensions of confidentiality, integrity and availability (CIA model) of the datasets.[77] Clarifications are thus needed to see how these dimensions can be integrated within the Blueprint.

Confidentiality is necessary to limit the access to private and personal (food consumption and background) information. Remarkably, the rationales underlying the confidentiality of these data are different: while the protection of commercially-sensitive information fosters the legitimate interests of private parties, the secrecy of personal data is necessary for reasons linked to fundamental rights to privacy and data protection. Security measures should be grounded on the conceptualisation of food consumption data as proxy variables for sensitive information. Unlawful processing activities may occur for illicit behaviours on legitimately processed data, e.g., beyond the "purpose limitation" principle. Individual food consumption data and background information might be used to predict the health status of individuals or for marketing purposes, hence the necessity of protecting this information from unlawful exploitation and data breaches.

As regards *integrity*, security should also be intended as the conceptual basis for safeguard measures that ensure that analysed data have not been altered. As remarked, data "poisoning" seriously threatens the validity of scientific findings and the credibility of all the stakeholders involved in food safety.[78] This dimension becomes relevant when we consider what has been remarked by AGID[79] when

---

74 Vedder (2019)

75 HLEG (2019), p. 17

76 BMWi (2018), p. 8

77 Olivier (2002)

78 This was already observed when discussing the Monsanto Paper scandal in Sect. 2.1.3

79 AGID (2018), p. 30

discussing the relationship between the security of machine learning systems and the construction of training datasets, according to the well-known mechanism 'garbage-in, garbage-out'.[80]

Finally, with regards to *availability*, security measures should not hinder data sharing among national food safety authorities, risk assessors and risk managers. When safeguards are necessary to ensure confidentiality, their design should allow legitimate uses of data for the validation of assessment results consistently with the new provisions of the Transparency Regulation, copyright and data protection law.

Discussing AI security *strictu sensu*, it can be observed that, in light of the lower degree of autonomy of machine learning systems in our domain in comparison to others (e.g., self-driving cars or robots), the relevance of technical robustness is quite limited. However, the dimension of error handling[81] shall not be overlooked, in particular when designing[82] systems used to support human decisions with large-scale effects.

5.2.2 *Accountability*

The principle of Accountability can be found in all the aforementioned ethical Charters. The German document[83] suggests that accountability frameworks are needed to generate trust towards machine learning systems, or our second-order trust. The Dutch document[84] takes a holistic approach to encompass three stages of AI development discussed hereafter.

In the *design* stage, accountability pairs with the replicability of the system's operations.[85] Ensuring this property is desirable in our domain since the validation of submitted or original studies cannot disregard the chance of experiment being replicated. When the studies rely on machine learning software, data and algorithms constitute part of the experimental setting and should favour its replicability to their maximum extent. This entails, *inter alia*, the necessity of identifying key elements

---

[80] In general, it has been observed that poor quality training data generate erroneous (e.g., biased, fallacious, or simply wrong) outputs. To understand the relevance of guaranteeing the integrity of data used in the training phases of AI development, the MIT carried out the "Norman" experiment (http://norman-ai.mit.edu, accessed on 27/04/2022). The Institute trained a conversational bot with different input data holding the same algorithm. Following training phases, the bot developed some sort of bipolar disorder: on the one hand, it displayed a regular interaction; on the other hand, it turned into a sociopath. Holding the training algorithm constant, the divergent aptitude of Norman reflects the quality of its training data

[81] European Commission (2020)

[82] BMWi (2018), p. 37

[83] BMWi (2018), p. 16

[84] SIGAI (2019)

[85] HLEG (2019), p. 17

of the mathematical model, the data labelling process, and other elements described under the "Explainability" principle.

In the *monitoring* stage, accountability implies the capability of verifying and audit the resources used in the course of risk assessment. This stage seems quite critical since EFSA's competences and financial resources might not suffice to ensure the monitoring of algorithms used by third parties when submitting scientific evidence or by the Authority itself. Case studies presented by EFSA[86] show that it can be performed with the support of machine learning systems, e.g., automatic literature reviews. Such techniques might also be performed over results generated with other machine learning approaches. Monitoring of algorithms seems then needed at every level of the scrutiny to prevent "cascade" accountability gaps.

Finally, in the *redress* stage, accountability is crucial to allocate liability for damage[87] or moral responsibility for the use of machine learning systems.[88] However, redress represents a legal and governance challenge. Elsewhere we noted (Sect. 4.2.2) that EFSA's scientific outputs are not subject to the jurisdiction of the CJEU due to their non-bindingness, Moreover, the remit of EFSA non-contractual liability is limited and never tested before a Court. Finally, let us introduce the hypothetical in which EFSA receives results based on a functioning (i.e., nonfaulty) machine learning algorithm that suffers from the probabilistic error typical of stochastic approaches.

A possible solutions relies on the consideration that the Commission is the entity entitled to issue binding decisions—informed by EFSA's scientific opinions—for which it holds accountability. Despite the critiques discussed in Sect. 4.1.2, the current regulatory framework holds accountable the risk manager only. When adopting decisions, the Commission should rely on the precautionary principle discussed in Sect. 4.2.3. Therefore, the principle can serve as a benchmark to allocate liability in the hypothetical: if probabilistic results cannot be guarantee a sufficient degree of certainty (e.g., a large range of plausible values in regression problems or a high number of false positives even in the case of high precision scores for classification algorithms), the precautionary principle is triggered. It follows that the most cautious measure (e.g., a temporary ban of a given product) is taken pending confirmatory results, perhaps with conventional methodologies. If this sort of mechanism is implemented, redress mechanisms for the use of machine learning techniques are reconciled with the consolidated redress scheme in the current legislation.

Mutatis mutandis, Accountability principle also governs the processing of personal data. On the one hand, this is true when individual food consumption data and background information are collected and stored since data controllers have to comply with data protection law;[89] on the other hand, even though the GDPR

86 IZSTO et al. (2017)

87 Villani et al. (2018), p. 113

88 SIGAI (2019), p. 5

89 Pagallo et al. (2019)

and Regulation 2018/1725 are not directly applicable to aggregated data, the close relationship between the information type at stake and personal (individual and group) identity and the consequences of the processing activities for certain groups entail the necessity of taking into account the context in which such processing occurs, even beyond the scope of data protection law.[90]

5.2.3 *Fairness*

From the premises highlighted in Sect. 4.1.1, fairness is a crucial principle intended to minimise the effects of possible discriminations, in particular of individuals belonging to groups that can be drawn on the basis of food preferences.

In the AI Charters under scrutiny, Fairness can be intended from a *procedural* (or design) and a *substantive* perspective. From a *procedural* point of view, it has been described as the technical prevention of biases in results generated through machine learning techniques.[91] Discriminatory outcomes can result from the poor quality of the datasets, their incompleteness or imbalances.[92] This can be avoided by adopting technical measures.[93] In its procedural dimension, fairness is also connected to accountability measures and redress mechanisms.[94]

From a *substantive* viewpoint, the use of machine learning techniques should be steered towards the minimisation of the negative consequences of existing discriminations and the accessibility of the benefits due to AI.[95] Substantive fairness also entails the prevention of new forms of wrongful differentiations.[96] As regards this last point, the considerations on data-driven fallacies Sect. 4.1.2 and the functioning of probabilistic modelling are of particular interest. Since discriminatory results can originate from functioning algorithms (i.e., algorithms that correctly execute all the planned instructions), assessing their unfair outputs might become even harder despite the availability of the source code and the training datasets.

In our domain, Fairness can then be discussed under both perspectives, which are inextricably linked. From a procedural standpoint, fairness can be understood as equal representativeness of food patterns. The datafication of food preferences and the use of dietary information for risk assessment purposes entails the risk that certain food patterns are under- or over- represented, thus leading to the data-driven fallacies discussed in Sect. 4.1.2. Such fallacies might generate discriminatory

---

[90] Vedder and Naudts (2017)

[91] Villani et al. (2018), 121

[92] European Commission (2020)

[93] SIGAI (2019), p. 11

[94] HLEG (2019), p. 7

[95] Jobin et al. (2019)

[96] Vedder and Naudts (2017)

results, in particular when imbalanced datasets are used as training sets for machine learning algorithms. Beside unfair differentiation based on "regular" data, such as geographical provenance or gender, discrimination can also occur on the basis of personal information which reflect individual traits subsumed under the category of sensitive data when food preferences are associated e.g., to ethnic or religious groups.

From a substantive perspective, the correspondence between food preferences and personal identity has been noted in Sect. 3.3. In the light of the above considerations, risk assessment shall not increase existing discriminations and shall trait all food patterns equally. Despite the lack of studies suggesting any perpetration of this form of discrimination, an explicit commitment of public and private entities to anti-discrimination measures is still missing despite the risks highlighted by the abundance of papers and documents on this topic, including institutional AI Charters. This is particularly relevant since non-traditional forms of discrimination—such as the ones based on vegetarianism or veganism—are yet unknown to anti-discrimination measures while being discussed in courts.[97]

Fairness can ultimately serve as the rationale for technical and organisational measures intended to prevent discrimination based on dietary patterns. Such measures can ensure that due consideration is given to non-traditional or minoritarian food preferences when they are recorded and used for further analysis (*procedural* perspective), according to the trend identified as the disparate impact of Big Data.[98] Such conceptualisation of *fairness-as-representativeness* is also grounded on the concept of statistical parity, i.e., the property that the demographics of those receiving positive (or negative) classifications are identical to the demographics of the population as a whole,[99,100] regardless of machine learning deployment. The reference to Rawls[101] in Dwork[102] makes explicit that the notion of fairness presented by the authors is individual rather than collective, hence the necessity of also evaluating its group-oriented dimension. Different solutions (namely, prejudicial remover techniques) have been proposed[103] and framed also considering the Explainability principle discussed below.[104] The most appropriate notion of fairness in our context seems to be a group-based conceptualisation that promotes equality

---

97 Case *Casamitjana v The League Against Cruel Sports* [2020] UKET 3331129/2018, discussed in Chap. 3

98 Barocas and Selbst (2016)

99 Dwork et al. (2012)

100 As the technical domain of food safety has been constructed both on classification and regression problems, further clarifications are needed. Despite the occurrence of regression problems in the domain at stake, statistical parity affects the binary *at-risk/not-at-risk* classification of aggregated food consumption patterns, regardless of how the problem has been framed (e.g., in the case of logistic regressions)

101 Rawls (2009)

102 Dwork et al. (2012)

103 Kamishima et al. (2012)

104 Zarsky (2016)

among food patterns that aggregate individual preferences reflecting identified clusters (vegetarians, vegans, ethnic minorities).

On the other hand, these measures support the effort to prevent forms of discrimination based on data that make explicit our personal identity when data protection law is not applicable due to data aggregation (*substantive* perspective). Moreover, if we consider the ongoing trends related to the use of *near* big data sources (e.g., mobile app-assisted self-recorded food consumption), the principle of Fairness can be linked to its conceptualisation under data protection law (discussed below) as a legal benchmark to verify that individual expectations of privacy are respected.

5.2.4 Explainability

Explainability is an ethical principle primarily intended to promote algorithmic transparency and prevent opaqueness and "black boxes". It has been endorsed by the Commission as a key technical requirement[105] functional to the evaluation of fairness.[106] The German approach[107] has identified its proximity to trust, as explainable machine learning models allows the assessment of legal compliance.[108] The corollary of Explicability has been proposed by AI4People group[109] and the HLEG,[110] which has referred to this principle as the capability of AI systems to communicate their operations and provide for a rationale for their output.

Consistently with our legal analysis, the domain of Explainability should be deemed different from the scope of Transparency *strictu sensu*. While the former concerns the ability to scrutinise the logic and the rationale underlying algorithmic results, the latter has been eventually conceptualised as a general attitude towards data disclosures for reasons of democratic oversight (Sect. 2.2). To avoid misconceptions and to promote consistency with food law terminology, it would be preferable to treat this principle separately.

As regards the principle of Explainability, the domain at stake provides food for thought. An EFSA-funded study[111] shows awareness towards the issue of algorithmic opaqueness and displays scores regarding the capability to assess the logic underlying their results. Two intertwined sets of claims can thus follow. On the one hand, that it is necessary to strike a balance between the degree of efficiency of machine learning techniques and the explainability of their results; on the other

[105] European Commission (2020)

[106] European Commission (2018)

[107] BMWi (2018), p. 16

[108] Doshi-Velez et al. (2017) and Mittelstadt et al. (2019)

[109] AI4People (2018)

[110] HLEG (2019)

[111] IZSTO et al. (2017)

hand, that the precautionary principle can serve as a benchmark similar to what has been discussed under the principle of Accountability and in Sect. 4.2.3.

Let us start from the first claim. Risk assessment concerns the discovery of unknown possible risks through scientific analysis. If the machine learning models deployed in this context are not auditable and their results are not explainable, it follows that EFSA will confront with a *ignotum per ignotius* situation, i.e., the scenario in which an explanation is more cumbersome than the phenomenon that the one it should clarify. This is not convenient in light of its goal of reducing uncertainty towards food-related risks, that is the ultimate objective of risk assessment.[112]

As regards the second claim, we already noted that the precautionary principle does not apply to risk assessment activities and only pertains to the area of risk management. EFSA scientific outputs *inform* risk managers, which in turn take decisions on the basis of the precautionary principle. Risk managers can allow for less cautious decisions only if the layers of uncertainty surrounding certain risks are erased to a threshold that is higher than the one needed to pass the precautionary test. When unexplainable machine learning models prevent a deep scrutiny of possible risks, the precautionary principle still applies. In other words, the opacity of machine learning models is an integral part of the scientific uncertainty that might lead to the "most cautious decision" to be taken by risk managers in accordance with the precautionary principle. Therefore, explainability and precaution are linked. Their relationship can be constructed as follows: the "most cautious decision" has to be taken every time the level of uncertainty due to the use of opaque machine learning algorithms is intolerably high according to a precautionary evaluation. When this is the case, a comparable conventional method should be preferred to verify the existence of risks. This indirect relationship also ensures a sufficient degree of consistency between Explainability and the existing legal framework, in particular as regards the accountability for the decisions that are made.

This is, of course, a hard trade-off. Viable alternatives may consist of researching how machine learning models deployed in this context can be made more explainable, consistently with the approach to explainable AI (XAI) research promoted by the French charter.[113]

The HLEG noted that the explanation "should be timely and adapted to the expertise of the stakeholder concerned. In addition, explanations of the degree to which an AI system influences and shapes the organisational decision-making process, design choices of the system, and the rationale for deploying it, should be available".[114] It is thus possible to conceive the essential traits of explanations. In general, explanations can be given *ex-ante* and *ex-post* and should be comprehensive

[112] From a broader perspective, the consideration of EFSA as an "autonomous epistemic agent" (see Lynch 2016) can reinforce the necessity of implementing explainable algorithms to increase its active participation to the epistemic aspects of risk assessment. In this perspective, explainable algorithms allow for the transformation of statistical correlations into epistemic conclusions by EFSA and remark its autonomy

[113] Villani et al. (2018)

of several elements identified by Palmirani (2020) and the Knowability approach when discussing explanations in automated decision-making (ADM) in the context of the GDPR. Explainable ADM targeted to individuals and explainability in our domain differ significantly due to the existence of specific norms granting a right to explanation (yet, debated among scholars[115]). However, the discussion on *what* would make algorithms explainable significantly affects this domain.

To prevent opaqueness of machine learning models applied to food safety risk assessment, essential elements[116] should be provided.[117] *Inter alia*, it is possible to lower the level of opaqueness by providing:

- *Ex-ante* elements:
  1. What mathematical models have been applied;
  2. How training, test and validation sets have been constructed (e.g., sources, missing data strategy, data cleaning strategy, measures to prevent biases, under- and over- fitting);
  3. How selection and grouping of individual food consumption data has occurred;
  4. If the individuals entitled to label data dispose of sufficient expertise to fulfil their tasks, in the case of supervised learning algorithms.
- *Ex-post* elements:
  1. An explanation of its outputs understandable for their receiver (Explicability, see below)
  2. A comparison between the results obtained with the support of machine learning techniques and conventional methods
  3. Overall performance scores (accuracy, recall, F1, etc.)
  4. Detailed performance scores (accuracy, recall, F1, etc.) for food patterns

Against these findings, it might be argued that providing these elements is an excessive burden for commercial applicants or the Authority. While this argument might seem *prima facie* acceptable, it is worth considering the centrality of trust in the scenario at stake. The submission and subsequent publication of these elements is ultimately aimed to reinforce trust and to prevent that machine learning models

---

[114] HLEG (2019), p. 18 and Doshi-Velez and Kim (2017) distinguish between *global* and *local* explanations. While global explanations refer to the general framework in which an automated decision has been taken (e.g., to detect biases), local explanaibility concerns the ability of giving reasons about a specific decision. See also Sovrano et al. (2019) and Sovrano et al. (2022)

[115] Wachter et al. (2017), Doshi-Velez et al. (2017), and Pagallo (2020)

[116] The list below has been drafted considering conclusions from Palmirani and Sapienza (2021) who split the *ex-ante* from the *ex-post* explainability elements, then adapted to the technical findings of this study. The list is also context-dependent, as some elements might be missing in practical scenarios

[117] These criteria should apply both in industry-to-EFSA, EFSA-to-Commission and EFSA-to-public information sharing depending on the entity that makes use of machine learning algorithms

put an additional layer of opacity and, thus, aversion towards the entities involved, including undertakings. On the one hand, the time span between these submissions and the renewal is sufficient to ensure return of investment both in financial and intangible assets (e.g., brand or institutional reputation); on the other hand, the submission of these elements is not revolutionary, as an effort is already made to submit and publish other information that results from considerable investments.

Explicability is a related ethical principle[118] first proposed by the AI4People initiative[119,120] that expresses the practice to select an appropriate Level of Abstraction that fulfils the desired explanatory purpose, appropriate to the system and receivers of the explanation, deploy suitable persuasive arguments and finally provides explanations to the receiver of its outputs on the goals pursues when developing and deploys the AI system.[121] Explicability becomes necessary every time AI outputs need to be understood and interpreted by their receiver. Its remit is not limited to the industry-to-EFSA data sharing, but also embraces the EFSA-to-Commission relationships and, crucially, the EFSA-to-citizen scenario, i.e risk communication.

Explanations shall not nudge the explainee to refrain from critically questioning or contesting the decision.[122] The instrumentality of explanations to the right to contest individual ADM has been noted by some authors.[123] In the domain under scrutiny, the right to contest the decision is not applicable. However, a conceptual substitute is EFSA's institutional duty to critically evaluate scientific evidence coming from commercial applicants. Instead, when the Authority relies on machine learning systems (e.g., automated literature review) Explainability is functional to the transparency and the accountability of risk assessors (Sect. 4.2). Therefore, Explainability significantly correlates with EFSA's mandate in both cases.

The nuance of Explicability also becomes relevant: while the targets of such explanations can be multiple, they undoubtedly include independent reviewers (e.g., NGOs) and the general public. Therefore, appropriate design strategies shall be designed to properly communicate these findings also to these elements without nudging the explainee towards certain interpretations.

---

[118] Jobin et al. (2019)

[119] AI4People (2018)

[120] AI4People group investigated two nuances of Explicability, namely "intelligibility" and "accountability", both of which have been kept into account in this chapter. See in Sect. 5.1.4

[121] Cowls et al. (2019)

[122] Mittelstadt et al. (2019)

[123] Palmirani (2020), Palmirani and Sapienza (2021), and Wachter et al. (2017)

5.2.5 *Transparency*

Thanks to the amendments brought by the Transparency Regulation, the EU food safety risk assessment legal framework has improved the capability of independent reviewers and authorities to carry out their activities by making more data available to them. At the same time, if the premises of the reform will be respected, all the stakeholders will eventually benefit from the gain in reputation and trust among citizens.

In the scrutinised AI Charters, Transparency has multiple facets. First, it encloses what has been previously defined as Explainability[124] when it empowers individuals to scrutinise the logic and the criteria underlying certain decision-making processes that make use of Big Data and machine learning.[125] While this dimension has already been covered and unnecessary repetitions might generate confusion, it is necessary to underline that Explainability and Transparency dimensions are both means to allow the intelligibility of the algorithmic results. What might differ, especially in the context under discussion, is their goal. While Explainability is mainly aimed to minimise the opaqueness of certain machine learning algorithms to increase the epistemic soundness of their outcomes, food law Transparency revolves around other obstacles that prevent the foreseeability of health-related risks (e.g., IPRs) or layers of opaqueness in the activities of the Authorities.

The notion of transparency has been defined holistically to comprise data, algorithms and business models[126] as a necessary prerequisite of auditability.[127] Importance has been given to the relationship between algorithmic transparency and the accountability of public bodies when they make use of AI systems.[128]

We noted that existing laws are more data-centric than algorithm-oriented. They seldom require the submission of programming codes used to perform analysis. A proactive and future-proof approach to Transparency might suggest implementing measures such as broadening the list of information to be disclosed by commercial applicants to include programming codes, input data and trained models to allow the replication of the experiment, or creating lists of algorithms and mathematical models that the industry has adopted when drafting submitted studies.

Let us now verify the consistency between this novel interpretation of transparency and the legal framework in place. When discussing the legal justifications of transparency, it has been noted that it consists of allowing citizens to foresee health-related concerns ("transparency-as-foreseeability") (Sect. 2.3) when environmental information is at stake and ensuring independence and accountability of the Authority (Sect. 4.2).

---

124 SIGAI (2019), p. 13

125 BMWi (2018), p. 38

126 HLEG (2019), p. 18

127 Villani et al. (2018), p. 15

128 AGID (2018), p. 11

It shall be remarked that the shift towards an algorithm-oriented model necessarily implies some constraints to the measures intended to protect the investments in R&D made by the industry, hence they should be limited to what is necessary and proportionate to their overall goal, which is worth examining.[129]

In both cases, a shift towards algorithms could be justified under the objectives of the existing regulatory framework. With regards to the independence and the accountability of the Authority, the release of details about the functioning of the algorithms deployed either by the industry or by the Authority itself is necessary to ensure that (a) EFSA is kept "in the loop" and its outputs do not entirely depend on algorithms, and (b) accountability gaps are prevented due to the absence of human oversight. When environmental information is at stake, the "transparency-as-foreseeability" criterion implies that details about algorithms are made available to the public to verify the correctness of their functioning and the findings related to human health that are generated with the support of these algorithms.

5.2.6 The Question of Privacy: From SAFETY to P-SAFETY

Finally, let us discuss the relationship between SAFETY principles and privacy. Since we identified a human informational component of food safety risk assessment and discussed its ownership implications, this subsection aims to cast light on what kind of privacy shall be discussed and how it relates to the SAFETY principles as a whole and with respect to each of them.

Both institutional and non-institutional AI Charters show unanimous consensus on privacy, hence the necessity of keeping it into account according to our methodology. It is conceived as the protection of the private sphere,[130] both at individual and group level.[131] When considered in its constitutional dimension in Germany, its role as a fundamental right is remarked.[132] With specific regard to machine learning techniques, due attention is given to the generation of training datasets that do not reflect biases or generate inaccuracies.[133]

As previously stated, AI Charters are purposively broad and embrace all the possible meanings of *privacy* in the sense given by Floridi,[134] i.e., intrusion in

[129] Instead of the "fostering trust" rationale, which is an original interpretation of this work, a safer starting point can be the reconstruction of the CJEU case law discussed above. This is also to avoid the *petitio principii* or circular reasoning fallacy. This Blueprint is intended to foster trust in all the entities involved in food safety risk assessment. Therefore, the "fostering trust" rationale underlying transparency measures cannot be used as their sole justification in this context, despite its validity in the construction of the ownership model discussed before

[130] AGID (2018)

[131] Villani et al. (2018)

[132] BMWi (2018)

[133] European Commission (2020)

[134] Floridi (2013), Chap. 12.2

physical, mental, decisional, and informational level. Mental and decisional privacy are not threatened using machine learning techniques in food safety risk assessment due to the lack of any direct intervention on individuals' mind or decisions. This includes individuals who are surveyed in food consumption data collection stages. While it is true that EFSA outputs can have some degree of influence over consumers, they do not primarily aim to nudge or manipulate individuals and their effects in orienting preferences seem quite limited without the support of risk communication. Moreover, the discussion on whether such manipulation is likely to occur falls under the scope of risk communication rather than risk assessment. The physical intrusion in the private sphere seems quite limited as well, especially in comparison to other technologies covered by institutional documents (e.g., facial recognition or AI-supported IoT devices). Quite different would have been the case had real-time food consumption data collection systems been the state-of-the art. Lastly, while informational privacy and data protection might have a strong relevance in data processing activities, technical and legal evidence discourages its explicit endorsement. As already discussed, food patterns are usually aggregated or otherwise anonymised, thus limiting the scope of data protection laws over processed data and the intrusion in individuals' private spheres derogations of data protection laws for statistical and research purposes consistently limit the remit of rights granted to protect 'raw' personal information.[135]

For these reasons, an outright adoption of Privacy as a standalone principle would be wrong and likely to raise methodological concerns. Therefore, the SAFETY model could adopt informational privacy as a meta-principle, i.e., a principle primarily intended to serve and integrate the SAFETY model. Such *meta*-formalisation of informational privacy does not entail that less importance should be given to it in comparison to other principles. Instead, this conceptualisation simply clarifies that privacy is included in the model "in relation to" (P-) the other principles rather than as an independent component. When informational privacy is conceptualised as data protection,[136] the legal framework also embeds "principles, general rights, concrete subjective rights and rules",[137] whose interplay with the Blueprint shall be taken into account. In the light of the limited applicability of data protection laws in the framework under scrutiny, the principle-based approach to privacy suggests that measures shall be taken *beyond* the applicable law and for reasons of trust. Once clarified the relational nature of the Privacy principle, let us now discuss how this relationship unfolds with respect to each component of the SAFETY model.

The connection between Privacy and Security has already been introduced when discussing the implications of data processing activities with regards to individuals' food consumption and background information. We noted that, as these data might be used as proxy variables to infer sensitive information about individuals, data

[135] Palmirani (2020)

[136] Lynskey (2014)

[137] De Hert (2017)

breaches and unlawful exploitations pose a serious threat to informational privacy. Alongside such consequentialist reasons, the ontological approach to privacy[138] widely discussed in Sect. 3.3 might suggests that violations to these data constitute a breach of one's personal identity. Such approach seems appropriate in light of the nature of the personal (food consumption) data at stake, hence the need of prioritising data protection measures for the personal information stored and analysed in this context. The same holds true for "AI security", i.e., the minimisation of threats posed to individuals by AI systems. In light of the limited degree of autonomy of machine learning systems in the domain at stake, Privacy relates with this facet of Security when considering that the use of machine learning techniques might generated inferences that re-identify individuals and possibly their sensitive traits.

Privacy can be a guidance to evaluate Accountability for the individual, collective and social implications of personal data processing. This can be done by two sets of measures: on the one hand, privacy-oriented accountability *strictu sensu* requires that efforts to comply with data protection law are made[139] and demonstrated.[140] They include technical and organisational measures ("data protection by design"), staff training and records of the data protection activities.[141] On the other hand, *latu sensu* accountability requires a risk-based aptitude towards the collective and social effects of large-scale data processing. While such formalisation of accountability may exceed the scope of data protection law, risks such as the possible re-identification or de-anonymization of surveyed individuals or data-driven biases towards certain groups identified by their 'datafied' food behaviours call data controllers to take responsibility for their data processing activities.

Likewise, the conceptualisation of Fairness could benefit from the Privacy-oriented consideration that food consumption data analysis can discriminate on grounds that consist of sensitive attributes of individuals. As we noted as regards Fairness, however, anti-discrimination law does not necessarily apply to all the grounds of discrimination that use food consumption data as proxies, hence the necessity of extending anti-discrimination measures beyond the material scope of anti-discrimination law. Hacker[142] argues that data protection law and anti-discrimination provisions can be integrated when personal data are used to make decisions to tackle discrimination issues. This might have implications regarding *indirect* discrimination—the only possible in the context at stake—and group-level privacy discussed in Sect. 3.2.3.[143] At individual level, when conceptualised as a

---

[138] Floridi (2005)

[139] Article 5(2) of the GDPR and Article 4 of Regulation 2018/1725 state that "[T]he controller shall be responsible for, and be able to demonstrate compliance with, paragraph 1 [*ed.* the other data protection principles]" See also De Hert (2017). In the context of data transmission, Art. 9(1) of Regulation 2018/1725 is also relevant.

[140] Instead, we noted elsewhere (Sect. 3.2) that the protection given to personal data at the collection stage is highly jeopardised

[141] ICO (2018)

[142] Hacker (2018)

data protection principle,[144] Fairness requires that personal data is processed in a way that does not harm data subjects. With the remit of automated decision-making being limited in food safety risk assessment, Fairness and Privacy work jointly to ensure that adverse effects on individuals are minimised. This might trigger measures that prevent the unlawful exploitation of their data for purposes other than risk assessment, limited access to data, and, from a broader perspective, that other data protection law principles are respected.[145] Regarding two of them—data minimisation[146] and accuracy[147]—trade-offs are necessary. On the one hand, data minimisation requires that the least possible amount of data is collected, hence the need of validating the information obtained from the data subject[148] (Sect. 3.1) in light of what data ought to be collected; on the other hand, in light of the limited remit of data subjects' rights, data accuracy requires an effort to be made to adapt recorded food consumption data to individual preferences over time. These two elements bridge the gap with machine learning techniques, which require a large quantity of up-to-date data to produce accurate results.

A large and growing number of scholars is debating on the relationship between data protection law and Explainability, especially in light of the provisions enshrined in Article 22 of the GDPR.[149] We noted that the remit of data protection law is restricted due to the statistical and research purposes of the processing for risk assessment and no automated decision-making process is made in the course of risk assessment. Regarding profiling,[150] it might be the case that certain data processing activities fall within the remit of predictions related to data subject's health when exposed to pathogens. The nature of data processing made by EFSA does not seem to cover this possibility. However, when considering industry-led studies that process dietary intake data with the support of predictive algorithms, the fate of inferred information might be relatively unknown to the data subject.[151] It is then

---

143 Floridi (2017)

144 Article 5(1) of the GDPR; Article 4(1) Regulation 2018/1725

145 Authors Sabelli and Tallacchini (2017) have underlined that such conceptualisation of "fairness" in data protection might be insufficient to ensure a fair data processing due to the *ex-ante* adoption of machine learning algorithms and the existence of *ex-post* remedies only in data protection law. Therefore, they called for a synergic inclusion of data protection law and collective values, including trust

146 Article 5(1)(c) of the GDPR; Article 4(1)(c) of Regulation 2018/1725

147 Article 5(1)(d) of the GDPR; Article 4(1)(d) of Regulation 2018/1725

148 See https://doi.org/10.5281/zenodo.1215993, accessed on 27/04/2022

149 See *inter alia* Wachter et al. (2017) and Palmirani (2020)

150 "Profiling" consists of any form of automated processing of personal data consisting of the use of personal data to evaluate certain personal aspects relating to a natural person, in particular to analyse or predict aspects concerning that natural person's performance at work, economic situation, health, personal preferences, interests, reliability, behaviour, location or movements" (Article 4(1)(4) of the GDPR, Article 3(1)(5) of Regulation 2018/1725). For an exhaustive review of GDPR's provisions with regards to profiling activities, including concrete cases, please refer to the work done by González and de Hert (2019)

necessary to clarify the extent to which individual predictions may occur and their effects on risk assessment activities. This is particularly true when considering the existence of "data conglomerates" such as the Bayer-Monsanto company which might use additional information to generate predictions based on food consumption data. When this is the case, data subjects should be given *ex-ante* and *ex-post* intelligible details about the data processing.

The relationship between Privacy and Transparency might be cumbersome to grasp. Transparency is a key principle of data protection law[152] as it contributes to the informational self-determination of the data subject by making her or him aware of the nature of the processing, the entities involved and so on, ultimately fostering her or his free choice. However, the meaning attributed to Transparency in the Blueprint—in essence, the widest possible availability of data and contextual information to replicate studies—differ from the one conferred in the context of data protection law. Therefore, when linked to the Blueprint, Privacy shall act a constraint to Transparency in giving access to raw personal information to third parties. This is necessary to prevent unlawful exploitations of data, including possible breaches of the purpose limitation principle.[153]

5.3 Possible Implementations of the P-SAFETY Blueprint

Our P-SAFETY Blueprint establishes principles for a trustworthy implementation of machine learning techniques in food safety risk assessment activities. The transition from 'data' to 'algorithms' might seem confusing at first. As argued in the Introduction, machine learning is progressively becoming the most popular approach for 'making sense' of data. Any merely data-centric set of recommendations would be outdated as soon as this computational approach becomes the dominant paradigm.

However, it is worth noticing that certain legal provisions are already in place. The Transparency Regulation revolves around data and sets rules that are obviously consistent with the Blueprint as it has been drafted bearing in mind the existing legislation. Therefore, a conceptual separation regarding what is not regulated (i.e., the use of machine learning algorithm in our context) from what is subject to legislation (i.e., data) has to be made. Therefore, subsection Sect. 5.3.2 will handle algorithmic implementation in the light of recent legislative proposals, whereas Sect. 5.3.3 will discuss how existing provisions can be re-interpreted in a manner

[151] While this is undoubtedly out of the scope of the EU food legislation, this consideration might be of interest for entities (including non-profit) and scholars that investigate the implications of personal data processing and its interplay with specific domain, e.g., the processing within the context of pharmaceuticals

[152] Article 5(1) of the GDPR; Article 4(1) Regulation 2018/1725

[153] Article 4(1)(b) of the GDPR; Article 4(1)(b) of Regulation 2018/1725

that is consistent with our Blueprint. Sect. 5.3.1 clarifies the relevance of our methodology, namely the middle-out approach.

5.3.1 *The Middle-Out Approach*

Before assessing how the Blueprint can be implemented in practice, it might be necessary to rely on some emerging methodological proposals, while keeping in mind that there is no established principle-to-practice translation method.[154] This subsection illustrates a possible unified methodological approach, the "middle-out" proposal.[155]

These authors have introduced the concepts of "middle- out layer" and "middle-out approach" to the debate on data, AI and web of data governance. Their investigation looks for "what lies in between" top-down and bottom-up approaches to legal governance. Traditional models consists of "top-down" (e.g., enforced regulation), and "bottom-up" (e.g., self-regulation). Authors' methodology aims to verify the extent to which the convergence between multiple regulatory systems—which include primary and secondary rules—strikes a balance between technology, market and social norms. While doing so, they keep into account principles—or middle-out interfaces—that steer various forms of regulation. Notably, the middle-out is theorised as a governance framework for data and AI, thus resulting appropriate for our twofold task.

Some words by the authors[156] are remarkable: "[t]he "middle-out" approach draws attention on how the balance is struck between multiple competing regulatory systems [...]. The model of governance for AI has to strike a balance between competing regulatory systems. The "middle-out" approach specifies what balance is remaining, once forms of top-down legislation and pure bottom-up solutions are discarded".

The Authors have adopted their approach when discussing *de iure condito* the GDPR and the principle of accountability, which was found to be GDPR's middle-out interface aiming to minimise risks related to data processing. In their view, Accountability lies in between GDPR principles and rules, on the one hand, and organisational and technical (by-design and by-default) measures, on the other.

First, this approach aims to strike a balance by multiple regulatory systems, consisting of primary and secondary rules.[157] As the normative discussion proposed in Chap. 2 has highlighted, several pieces of legislation concur to define the legal framework of interest: a general regulation (the GFLR), sectoral legislation, and other provisions that relate to specific portions of the data processing activities

154 Mittelstadt et al. (2019)

155 Pagallo et al. (2019)

156 Pagallo et al. (2019)

157 Hart (1961)

at stake (e.g., the processing of personal data and the GDPR). Consistently with our data-centric approach, the middle-out layer of analysis should be focused on data-related provisions. Moreover, as the Authors remark, "the law is not the only regulatory system out there": market practices, social values and principles and even technology[158] shall be conceived as regulatory systems.

Secondly, the middle-out approach investigates and verifies the alignment between primary and secondary rules of the law. In our domain, normative implementations are left to EFSA's discretion when governing data from a legal (e.g., EFSA's confidentiality claims handling) or technical (e.g., data standardisation, additional data requirements) perspective. The middle-out layer of analysis allows us to identify at least a common denominator among these vertical normative sources when applied to the domain at stake.

Thirdly and most crucially, this approach seeks to coordinate bottom-up and top-down regulatory options. As regards the latter, market practices, principles and technology can govern human behaviour and possibly be translated into policy choices. The middle-out layer approach allows us to frame their relationship to be appropriate to our domain. For instance, we noted how principles of openness and transparency are embedded into the system and reflected into the new transparency measures endorsed by the amendments to the GFLR.

To this end, our Blueprint constitutes one of the many possible implementations of any data governance framework applicable to the domain of food safety that is based on promoting trust, thus including the Transparency Regulation and the complementary legislation. In doing so, it is agnostic to the nature of the regulatory model (regulation, self-regulation, co-regulation, etc.) while preserving the necessary flexibility to be implemented in any policy option.

5.3.2 Hard Law, Self-Regulation, Codes of Conduct

Authors have noted that 'Translating principles into practice and seeking harmonization between AI ethics codes (soft law) and legislation (hard law) are important next steps for the global community'.[159] Policy options for the implementation of the Blueprint can range across various models identified by the literature.[160] Among them, three options will be discussed, namely 'hard law', self-regulation, and codes of conduct.

First, 'hard law' is defined as a complete legal framework that details binding obligations, usually subject to punishments and fines for non-compliance.[161] While

[158] Lessig (2009)

[159] Jobin et al. (2019)

[160] Petit and De Cooman (2021)

[161] Abbott and Snidal (2000)

obligations should be sufficiently precise, Courts' intervention would be needed to interpret them.

Would a 'hard law' implementation of the Blueprint be possible? Yes, but it is unlikely to happen in the near future, at least in the domain of food safety law. In the recent years (2018–2021), food law and policymakers have focused on data transparency, security, and confidentiality. Even though these legislative efforts are commendable for their timely intervention right after data-driven scandals, machine learning algorithms have been left without clear guidance on their use. Considering that the first complete reform of the GFLR with some relevance for ICTs has come after nearly 20 years from its original adoption, it is unlikely that any further modification will be approved in the near future.

Then, one could argue that the right hard law implementation of the Blueprint would be the foreseeable AI Act.[162] The regulatory model of the AI Act does not seem to endorse a strict 'hard law' approach. Authors have discussed a model of 'monitored self-regulation' for AI[163] which seems to be mirrored by the AI Act, at least for what concerns high risk AI applications. In this model, 'hard law' establishes a general framework with minimum requirements, whereas the implementation is left to the undertakings. *Ex post* mechanisms, including audits and sanctions, ensure compliance with the essential framework. However, it is worth reminding that "the choice between hard law and soft law is not a binary one".[164]

As regards the contents of the Act, it has been noted that machine learning applications described in this book do not fall within prohibited or high risk AI, despite being subject to the Act.[165] The Exploratory Memorandum of the proposal further clarifies that provisions of Title IV (limited risk) applications are those concerning manipulative AI systems. Transparency obligations in this Title will apply for systems that interact with physical persons, are used to detect emotions, or determine association with (social) categories based on biometric data, or generate or manipulate content ('deep fakes'). AI applications in food safety might fall within the first category if a (very) extensive interpretation of 'interaction with humans' is pursued. However, considering Art. 52 of the Proposal, it might be the case that 'interaction' is interpreted as a constant loop mechanism between human and machine behaviours. In this scenario, the 'interaction' terminates after the first iteration.

---

[162] Please note that, at the time of writing, the AI Act is still in the early stages of its discussion. This entails that further modifications to the Act during the legislative drafting might considerably alter the validity of the following statements

[163] Renda et al. (2019)

[164] Abbott and Snidal (2000)

[165] Current version of Annex III labels as 'high risk' AI systems those used in the contexts of (1) Biometric identification and categorisation of natural persons; (2) Management and operation of critical infrastructure; (3) Education and vocational training; (4) Employment, workers management and access to self-employment; (5) Access to and enjoyment of essential private services and public services and benefits; (6) Law enforcement; (7) Migration, asylum and border control management; (8) Administration of justice and democratic processes

Finally, title IX creates a framework for the creation of codes of conduct for providers of non-high-risk AI systems. These voluntary mechanisms embrace the provisions set for high-risk AI systems and broaden them to include other kinds of AIs. The Explanatory Memorandum clarifies that '[t]hose codes may also include voluntary commitments related, for example, to environmental sustainability, accessibility for persons with disability, stakeholders' participation in the design and development of AI systems, and diversity of development teams'. This seems closer, together with provisions regulating AI sandboxes (Art. 55), to the idea of 'adaptive regulation' discussed by some authors.[166]

Considering the large-scale impact for individuals and groups, the externalities produced by machine learning applications are subject solely to low-burden obligations of transparency. They should require—at least—the implementation of codes of conducts given their potential societal and collective impact. This does not imply that other implementations would not be possible, but they are less likely to occur in the short term in the light of the current legislative framework.

Code of conducts by the food industry revolve around data transparency. In 2018, CropLife International and the European Crop Protection Association, two associations representing companies active in the field of plant protection products, have released a "Global Transparency Initiative".[167] It might been questioned whether this was an attempt to escape from possible legislative intervention in the EU, given the time frame, or the a genuine attempt to foster education against supposedly misleading against their products, as it has been argued.[168] As regards EFSA, policy papers are very common[169] and they could also represent a code of conduct under the AI Act.

Let us identify some possible contents for such voluntary self-regulation consistently with the trust-oriented and principle-based approach taken by the Blueprint:

- Ensuring that algorithmic output are not biased or polarised towards certain groups, e.g., by ensuring equal representativeness of all food patterns and dietary habits, in particular those corresponding to known grounds of discrimination;
- Assessing and clarifying the role of "transparency scores" of machine learning algorithms;
- Developing cross-validation mechanisms that compare results determined by machine learning algorithms and traditional deterministic methodologies;
- Releasing information on the performance scores of the algorithms used in the course of risk assessment (see the detailed list at Sect. 5.2.4 on Explainability);
- Publishing information on the mathematical structure, source codes and log files of the algorithms used by the industry to perform their studies;

[166] Pagallo et al. (2019)

[167] Available at https://croplifeeurope.eu/media/press-releases/press-release-industry-data-transparency/, accessed on 27/04/2022

[168] Blake (2019)

[169] See EFSA (2015) and Cappè et al. (2019)

- Releasing synthetic data not covered by intellectual property protection. These data could be used to perform independent reviews—with or without machine learning—by using freely available information;
- Adopting data protection protocols capable of determining whether data are sufficiently anonymised/aggregated to prevent inferences capable of re-identify individuals;
- Adopting data protection protocols capable of determining whether data do not allow inferences revealing sensitive information of individuals.

5.3.3 Case-Law and Judicial Interpretation

Possible implementations of the Blueprint might also include judicial interpretations by European courts. As shown in Sect. 2.2.3, the European case law on data related to food safety is notable and recent cases have brought new attention on this subject matter. Our Blueprint can contribute to this debate on at least two main issues: information disclosure by EU institutions and the interpretation of the precautionary principle, in line with the competences of the CJEU.

On the one hand, the concept of 'environmental information' has been discussed in the *Greenpace and PAN Europe* and *Bayer CropScience* cases (see Sect. 2.2.3). In summary, the Court concluded that information on current and foreseeable emissions in into the environment shall be released as a matter of public interest (*Greeenpace and PAN Europe*, paras 79–80). Notably, it consists of data concerning the medium to long-term consequences of emissions, measured in realistic conditions, yet in 'most unfavourable' testing scenarios (*Bayer CropScience*, para 91).

The joint reading of Explainability and Transparency in our P-SAFETY Blueprint suggests the possibility of a novel interpretation of such information disclosure. In particular, the data-*centrism* of the Court has to be noted: while considering raw data necessary to understand the long-term implications of releases into the environment, the way in which such forecasts are drawn is not subject to the disclosure obligations.

It is important to recall here that our interpretation of data ownership (Sect. 2.3.4) is a dynamic one: each actor enjoys a limited account of ownership and information transmissions are needed to reinforce trust. If only data on emissions are released while no information on the algorithms is released, can there be trust?

Literature on trust in technology (see Sect. 5.1.1) suggests a negative reply. When such projections are drawn by means of machine learning algorithms, the release of source codes might not be sufficient and other elements should be disclosed for Explainability reasons (see the detailed list at Sect. 5.2.4).

This conceptual operations requires a novel interpretation of the material scope of the Transparency principle and the endorsement of Explainability when machine learning techniques constitute an integral part of the testing scenario. This interpretation does not seem unlikely. As discussed above (Sect. 2.2.3), the ECJ adopts a principle-based approach that first conceptualises principles to be adopted and then decides what information must be released accordingly. While Transparency is already a cornerstone of the EU food safety system and its adaptation to machine learning seems feasible, Explainability has still to be included. However, in the light of ongoing legislative actions, a progressive acknowledgement of its role in public decision-making might lead the ECJ to consider it applicable to our domain.

On the other hand, Explainability also relates to the second possible avenue of judicial implementation of the Blueprint with regards to the precautionary principle. In the previous sections I have introduced this principle in Sect. 4.2.3 and discussed its importance within the Explainability framework in Sect. 5.2.4. To briefly summarise these preliminary findings, machine learning algorithmic results always embed a statistical error that has to be kept into account when assessing their certainty. Moreover, their opaqueness shall constitute part of the scientific uncertainty surrounding certain results. Both these findings have implications as regards the precautionary principle and on the concept of uncertainty.

Accountability and liability frameworks can thus be revised to be appropriate to handle machine learning results. We noted in Sect. 4.2.2 that EFSA's opinions are becoming substantially binding and this is not desirable due to the formal lack of regulatory power and the substantial lack of judicial scrutinisability of their assessment. These issues can be exacerbated by the use of machine learning algorithms that serve as 'oracles' for scientific assessments that become *de facto* valid as binding decisions. The aforementioned peculiarities of machine learning algorithms call for an additional analysis of their outputs as regards the level of certainty of the results. Such analysis can solely be performed by risk managers, whose tasks include the balance between the performance scores of the AI system against the level of uncertainty of the results, which include their opaqueness. To put it differently, risk managers enjoy a greater higher margin of appreciation for the decisions taken on assessments performed by machine learning algorithms in comparison to traditional methodologies due to the higher number of factors (confidence scores, replicability, opaqueness, availability of source codes) that they have to keep into account when assessing the reliability of the results. Such assessment is necessary in the light of the precautionary principle and necessary entails a novel and less data-centric interpretation of scientific uncertainty.

5.4 Chapter Synopsis

Chapter 5 has presented the P-SAFETY Blueprint, a principle-based set of recommendations aimed at steering the adoption of advanced data analysis techniques, including machine learning, in food safety towards certain principles.

The chapter has covered a wide range of topics. In particular, its main aims were to provide answers to the questions left unanswered in the previous chapters and contribute to the ongoing debate on AI ethics by offering a significant case study on a relatively overlooked field of study.

This chapter has first introduced the goals and the conceptual background of the Blueprint. It consists of a short and concise document that can be used to interpret and align the existing data ownership and data governance frameworks to the challenges posed by Big Data collection, storage, and analysis, including by means of machine learning The Blueprint is presented as a non-hierarchical and non-exclusive list of principles that, taken together, can help EFSA and the industry involved to increase their trustworthiness by managing their data according to certain *desiderata*. The background consists of regulatory initiatives in the EU and, considering the ongoing legislative process, supporting documents such as reports by the EU Commission, Member States and interdisciplinary panels.

Then, the presented Blueprint consists of five core principles—Security, Accountability, Fairness, Explainability, Transparency (SAFETY)—and one metaprinciple—Privacy (P-)—that can be implemented to adapt the technical and legal domain of food safety risk assessment to the ongoing datafication trend in a manner that risks are prevented, and benefits are maximised. This is consistent with the current regulatory approach in the EU. These principles show adherence to the background documents while being adapted to the context of food safety risk assessment.

Finally, some possible implementations are discussed. It is necessary to keep in mind some come caution is needed in the evaluation of these proposals: the current legislative framework will likely be subject to amendments in the near future. Nonetheless, some proposals have been made. They include, consistently with the conclusions reached in previous chapters, information disclosures, methods of cross-validation of scientific evidence, releases of (synthetic) data, data protection protocols, and so on. Perhaps, the most viable and flexible way of implementation will be self-regulation. With the development of the legislative framework, new adaptations of old principles (e.g., transparency) can allow the Blueprint to be implemented via judicial interpretation.

References

Abbott KW, Snidal D (2000) Hard and soft law in international governance. Int Organ 54(3):421–456

AGID, Agenzia per L'Italia Digitale (2018) Libro Bianco sull'Intelligenza Artificiale al servizio del cittadino. AGID, Rome

AI4People (2018) AI4People | Atomium. https://www.eismd.eu/ai4people/

Alexy R (2000) On the structure of legal principles. Ratio Juris 13(3):294–304

Alexy R (2003) On balancing and subsumption. A structural comparison. Ratio Juris 16(4): 433–449

Barocas S, Selbst AD (2016) Big data's disparate impact. Calif. L. Rev 104:671

Beauchamp TL, Childress JF et al. (2001) Principles of biomedical ethics. Oxford University Press, Oxford

Blake R (2019) Is the european crop protection industry threatened by the drive towards transparency? Outlooks Pest Manag 30(2):50–52

BMWi, German Federal Ministry for Economic AffairsandEnergy (2018) Artificial intelligence strategy. https://www.ki-strategie-deutschland.de/home.html?file=files/downloads/Nationale_KI-Strategie_engl.pdf. Accessed 26 April 2022
Cappè S, Gilsenan M, O'Dea E, Richardson J, Verloo D (2019) The future of data in EFSA. EFSA J 17(1):e17011
Cowls J, King T, Taddeo M, Floridi L (2019) Designing AI for social good: Seven essential factors. Available at SSRN 3388669.
De Hert P (2017) Data protection as bundles of principles, general rights, concrete subjective rights and rules: piercing the veil of stability surrounding the principles of data protection. Eur Data Prot L Rev 3:160–179
Doshi-Velez F, Kim B (2017) Towards a rigorous science of interpretable machine learning. Preprint arXiv:1702.08608
Doshi-Velez F, Kortz M, Budish R, Bavitz C, Gershman S, O'Brien D, Schieber S, Waldo J, Weinberger D, Wood A (2017) Accountability of AI under the law: The role of explanation. Preprint arXiv:1711.01134
Durante M (2010) What is the model of trust for multi-agent systems? Whether or not e-trust applies to autonomous agents. Knowl Technol Policy 23(3–4):347–366
Dwork C, Hardt M, Pitassi T, Omer R, Zemel R (2012) Fair ness through awareness. In: Proceedings of the 3rd innovations in theoretical computer science conference, pp 214–226
EFSA (2015) Principles and process for dealing with data and evidence in scientific assessments. EFSA J 13(6):4121
European Commission (2018) Communication artificial intelligence for Europe. https://ec.europa.eu/digital-single-market/en/news/communication-artificial-intelligence-europe. Accessed 26 April 2022
European Commission (2020) Commission white paper on artificial intelligence—A European approach to excellence and trust. https://ec.europa.eu/info/sites/info/files/commission-white-paper-artificial-intelligence-feb2020_en.pdf. Accessed 26 April 2022
Feteris ET (2017) Fundamentals of legal argumentation, vol 1. Springer, Berlin
Floridi L (2005) The ontological interpretation of informational privacy. Ethics Infor Technol 7(4):185–200
Floridi L (2008) Artificial intelligence's new frontier: Artificial companions and the fourth revolution. Metaphilosophy 39(4–5):651–655
Floridi L (2013) The ethics of information. Oxford University Press, Oxford
Floridi L (2017) Group privacy: A defence and an interpretation. In: Group privacy. Springer, Berlin, pp 83–100
Floridi L, Cowls J, Beltrametti M, Chatila R, Chazerand P, Dignum V, Luetge C, Madelin R, Pagallo U, Rossi F et al. (2018) AI4People—An ethical framework for a good AI society: opportunities, risks, principles, and recommendations. Minds Mach. 28(4):689–707
Future of Life Institute (2017) Principles developed in conjunction with the 2017 Asilomar conference. In: Principles developed in conjunction with the 2017 Asilomar conference [Benevolent AI 2017]
Gellert R (2018) Understanding the notion of risk in the general data protection regulation. Comput Law Secur Rev 34(2):279–288
González EG, de Hert P (2019) Understanding the legal provisions that allow processing and profiling of personal data ? an analysis of GDPR provisions and principles. In: Era forum, vol 19, 4th edn. Springer, Berlin, pp 597–621
Grodzinsky FS, Miller KW, Wolf MJ (2011) Developing artificial agents worthy of trust: Would you buy a used car from this artificial agent? Ethics Inf Technol 13(1):17–27
Hacker P (2018) Teaching fairness to artificial intelligence: Existing and novel strategies against algorithmic discrimination under EU law. Common Market Law Rev 55(4):1143–1185
Hart HLA (1961) The concept of law. Oxford University Press, Oxford
HLEG, High Level Expert Group on AI (2019) Ethics guidelines for trustworthy AI. Available at https://digital-strategy.ec.europa.eu/en/library/ethics-guidelines-trustworthy-ai. Accessed on 27/04/2022

ICO (2018) Accountability and governance. https://ico.org.uk/for-organisations/guide-to-data-protection/guide-to-the-general-data-protection-regulation-gdpr/accountability-and-governance/. Accessed 26 April 2022
IEEE (2017) Ethically aligned design. IEEE Standards v2. IEEE, Piscataway
IZSTO, Ru G, Crescio MI, Ingravalle F, Maurella C, UBESP, Gregori D, Lanera C, Azzolina D, Lorenzoni G et al. (2017) Machine learning techniques applied in risk assessment related to food safety. EFSA Support Publ 14(7):1254E
Jobin A, Ienca M, Vayena E (2019) The global landscape of AI ethics guidelines. Nat Mach Intell 1(9):389–399
Kamishima T, Akaho S, Asoh H, Sakuma J (2012) Considerations on fairness-aware data mining. In: 2012 IEEE 12th international conference on data mining workshops. IEEE, Piscataway, pp 378–385
Kuner C, Cate FH, Lynskey O, Millard C, Ni Loideain N, Svantesson DJB (2018) Expanding the artificial intelligence-data protection debate. Int Data Privacy Law 8(4): 289–292
Lauritsen M (2015) On balance. Artif Intell Law 23(1):23–42
Lessig L (2009) Code: And other laws of cyberspace. Basic Books, New York
Lynch MP (2016) The internet of us: Knowing more and understanding less in the age of big data. WW Norton & Company, New York
Lynskey O (2014) Deconstructing data protection: The added-value of a right to data protection in the EU legal order. Int Comp. LQ 63:569
Mittelstadt B (2019) AI ethics–too principled to fail? arXiv preprint arXiv:1906.06668
Mittelstadt B, Allo P, Taddeo M, Wachter S, Floridi L (2016) The ethics of algorithms: Mapping the debate. Big Data Soc. 3(2):2053951716679679
Mittelstadt B, Russell C, Wachter S (2019) Explaining explanations in AI. In: FAT 2019: Proceedings of the conference on fairness, accountability, and transparency, pp 279–288
Olivier MS (2002) Database privacy: balancing confidentiality integrity and availability. ACM SIGKDD Explorations News Lett. 4(2):20–27
Pagallo U (2020) Algoritmi e conoscibilità. Rivista di Filosofia del Diritto 9(1):93–106
Pagallo U, Aurucci P, Casanovas P, Chatila R, Chazerand P, Dignum V, Luetge C, Madelin R, Schafer B, Valcke P (2019) AI4People-on good AI governance: 14 priority actions, a SMART model of governance, and a regulatory toolbox. Available at SSRN: https://ssrn.com/abstract=3486508. Accessed on 27/04/2022
Pagallo U, Casanovas P, Madelin R (2019) The middle-out approach: assessing models of legal governance in data protection, artificial intelligence, and the Web of Data. Theory Pract Legis 7(1):1–25
Palmirani M (2020) Big data e conoscenza. Rivista di Filosofia del Diritto 9(1):73–92
Palmirani M, Sapienza S (2021) Big data, explanations and knowability ragion pratica. Ragion Pratica (2):349–364
Pavon J, Gonzalez-Espejo MJ (2020) An introductory guide to artificial intelligence for legal professionals. Wolters Kluwer, Alphen aan den Rijn. ISBN:9789403509822
Petit N, De Cooman J (2021) Models of law and regulation for AI. The Routledge Social Science Handbook of AI. Routledge, London, pp 199–221
Quintarelli S, Corea F, Fossa F, Loreggia A, Sapienza S (2019) Una prospettiva etica sull'Intelligenza Artificiale: principi, diritti e raccomandazioni. BioLaw (3):183–204
Rawls J (2009) A theory of justice. Harvard University Press, Cambridge
Renda A et al. (2019) Artificial intelligence. CEPS Centre for European Policy Studies, Brussels
Russell S, Norvig P (2010) Artificial intelligence: a modern approach. Prentice Hall, Hoboken
Ryan M, Stahl BC (2020) Artificial intelligence ethics guidelines for developers and users: clarifying their content and normative implications. J Inf Commun Ethics Soc
Sabelli C, Tallacchini M (2017) From privacy to algorithms? fairness. In: IFIP international summer school on privacy and identity management. Springer, Berlin, pp 86–110
Sachs G (2016) Introduction to European food law and regulation. In: International food law and policy. Springer, Berlin, pp 409–450

Sartor G (2010) Doing justice to rights and values: teleological reasoning and proportionality. Artif Intell Law 18(2):175–215

SIGAI, The Special Interest Group of AI (2019) Dutch AI Manifesto. http://ii.tudelft.nl/bnvki/wp-content/uploads/2019/09/Dutch-AI-Manifesto-2019.pdf. Accessed 26 April 2022

Sovrano F, Vitali F, Palmirani M (2019) The difference between explainable and explaining: requirements and challenges under the GDPR. XAILA @JURIX 2019. Available at http://ceur-ws.org/Vol-2681/xaila2019-paper1.pdf. Accessed on 27/04/2022

Sovrano F, Sapienza S, Palmirani M, Vitali F (2022) Metrics, explainability and the European AI act proposal. Multidiscip Res J 5(1):126–138

Taddeo M (2010) Modelling trust in artificial agents, a first step toward the analysis of e-trust. Minds Mach 20(2):243–257

Taddeo M, Floridi L (2018) How AI can be a force for good. Science 361(6404):751–752

Veale M, Borgesius FZ (2021) Demystifying the Draft EU Artificial Intelligence Act. Preprint arXiv:2107.03721

Vedder A (2019) Safety security and ethics. Anton vedder safety security and ethics. In: Anton V, Jessica S, Charlotte D, Peggy V (eds) Security and law legal and ethical aspects of public security cyber security and critical infrastructure security. Intersentia, Cambridge/Antwerp/Chicago, pp 11–26

Vedder A, Naudts L (2017) Accountability for the use of algorithms in a big data environment. Int Rev Law Comput Technol 31(2):206–224

Verheij B (2020) Artificial intelligence as law. Artif Intell Law 28(2):181–206

Villani C, Bonnet Y, Rondepierre B et al. (2018) For a meaningful artificial intelligence: Towards a French and European strategy. Conseil national du numérique.

Wachter S, Mittelstadt B, Floridi L (2017) Why a right to explanation of automated decision-making does not exist in the general data protection regulation. Int Data Privacy Law 7(2):76–99

Wachter S, Mittelstadt B, Russell C (2017) Counterfactual explanations without opening the black box: Automated decisions and the GDPR. Harv J Law Technol 31:841

Wagner B (2018) Ethics as an escape from regulation: From ethics-washing to ethics-shopping. Amsterdam University Press, Amsterdam

Zarsky T (2016) The trouble with algorithmic decisions: An analytic road map to examine efficiency and fairness in automated and opaque decision making. Sci Technol Hum Values 41(1):118–132

Chapter 6
Conclusion: A Responsible Food Innovation

Let us discuss some final remarks on the wide range of topics discussed in this book. This conclusive chapter serves both as a general summary of our main findings and as the identification of the many questions that have been left open.

This study consists of an in-depth look into an often under-estimated—yet, highly debated—field of study, namely food safety. In particular, it revolves around one of its most controversial sub-domains, i.e., data used for risk assessment purposes. Hence, this contribution aims to foster the debate surrounding a research area to which the scientific literature on Big Data, ownership, and governance has not devoted adequate attention, especially in comparison to 'neighbour' subject matters such as healthcare and pharmaceutical.

Food safety could then become of prime importance. It consists of a unique ecosystem in which several data-types—non-personal, personal, mixed, observed, derived, inferred, and so on—co-exist. Such concurrence raises questions regarding the level of access to these data, the protection of intellectual property, the public relevance of such information, and so forth. This contribution also casts light on some of these issues, with the hope that other contributors will accept the challenges raised by this domain.

Among these information-types, food consumption data present unique challenges widely discussed in this study. Not only do they revolve around the scope of data protection law, but also their use in aggregated forms might raise concerns that go beyond the remit of the GDPR and Regulation 2018/1725. In both cases, this work contributes to these lines of research with an applied case study highlighting the need to synergise the debates on data protection law and other data-related concerns, including group privacy and discrimination.

Let us discuss in detail our findings while posing new questions for further thinking.

S. Sapienza, *Big Data, Algorithms and Food Safety*, Law, Governance and Technology Series 52, https://doi.org/10.1007/978-3-031-09367-8_6

6.1 Restoring Information Asymmetries

A primary area of investigation of this book is data ownership. The issue has been presented as the presence of conflicting interests between the actors involved in food safety (the industry, the competent authorities, citizens, etc.), which results in the lack of a shared answer as regards the question 'who owns the data?'.

The lack of a clear solution has been justified by the several pieces of legislation that attribute faculties to all the actors involved: while the industry can enjoy ownership on their data via IPRs or contracts, citizens' access to the same information is guaranteed by other pieces of legislation. The balance is struck by EFSA and the European Commission, under the jurisdiction of the ECJ, which must carefully consider the interplay of rather complex and occasionally cumbersome legal rules.

From a broader governance perspective, striking this balance can also be read as an attempt to restore an information asymmetry between the industry and the citizens to foster trust. Food safety information plays a key role in public decision-making in the area of public health. Trust become essential in supporting data-driven decision making.

As a matter of fact, the industry generates and accesses large *corpora* of information regarding their products and justifies certain claims on their basis. Conversely, access to this information is restricted to the citizens not only due to rules preventing them to process these data, but also due to their lack of skills and time to verify them.

Analysing this information is nearly impossible for common individuals due to their lack of skill and judgement. Therefore, one could argue that, due to the lack of effective contributions by the citizens, letting them access data to cross-validate scientific results is not necessary.

I would disagree with this claim. With the ongoing 'datafication' of food safety risk assessment and the increasing relevance of digital information, public availability of scientific data becomes necessary to restore the informational imbalance between the industry and citizens, those 'who know' and those 'who are willing to know'. We have observed that this has been the rationale underlying access to environmental information and we proposed a 'dynamic ownership' model to explain and justify how this balance can be struck for trust reasons.

Qualified intermediaries should be empowered to effectively and access data for cross-validation purposes and share their findings with the citizens. EFSA could provide guidance and rules to begin this process. Its institutional role seems appropriate to be equally distant from all the involved stakeholders interested in accessing this information. Data analysis should take place free from ideological and technical biases. Its findings should be communicated to the public accordingly. It is also time to extend the rationale underlying data disclosure to other informational components, i.e., algorithms used to gather scientific evidence. This book has stressed their growing importance, which should be translated into disclosure framework that allow the way 'we make sense of data' to be accessed and communicated.

Dynamic ownership is meant to restore informational imbalance. This does not entail that access to private information shall be unlimited. This is the case with data that constitute informational assets for companies. Derogations to the rules preserving these assets—while interpreted broadly—should be limited to purposes such as scientific research. Such limitations also concern the nature of the accessible data, which should be the ones that effectively contribute to the cross validation. Likewise, dynamic ownership entails that access to personal information shall be limited and balanced with the exigence of accessing food safety data. Throughout the book we provided sufficient justifications to this necessary safeguard for fundamental rights and freedoms. In practice, the proposed Data Governance Act and the 'data altruism' schemes could be a viable way for this implementation. Yet, the ongoing legislative debate makes hard to foresee the extent to which this hypothesis would turn into reality.

6.2 Food Safety Authorities, the Industry and AI: Joint Efforts Towards Trustworthiness

Trust is a key element of public decision-making in the area of food safety. This is because the long-term effects of food-related products are not easily foreseeable and only trusted entities can remove the layers of uncertainty surrounding certain hazards. Machine learning algorithms and artificial intelligence represent novel trends in risk assessment.

Trust be improved by achieving more transparency in risk assessment by disclosing the information on which it is based, but it is also necessary to provide forward-thinking solutions that can maintain a high level of confidence in the systems also when machine learning will become the state of the art in risk assessment.

With regards to trustworthiness, this study also strengthens the centrality of trust. While the aforementioned studies have identified the intrinsic necessity of Trust in human-machine relationships, the scenario discussed in this research also presents an external need of a fiduciary liaison that is independent from the use of AI/machine learning techniques, i.e. what has been referred to as 'First-order Trust'. However, the presence of a human-machine trust-based relationship (our 'Second-order Trust') reveals the existence of a new order ('Third-order Trust') which consists of the connection between citizens and the use of advanced data analytics techniques. While this is somewhat similar to the Second-order for being a human-to-machine relationship, Third-order Trust also shows the presence of an authoritative intermediary (EFSA).

In general, we noted that algorithms are placed outside the legal discourse. While this is justifiable by the conflicting interests of the involved stakeholders, this is not desirable in the light of a trust-oriented, future-proof discourse. To date, committing to self-voluntary releases of operational information on algorithms used in this context seems the only viable way for the industry. Codes of conducts are explicitly proposed and endorsed by the forthcoming AI Act.

R&D activities on the industry should be steered towards achieving trustworthiness by guaranteeing that the minimal requirements expressed in the P-SAFETY model and its theorised implementation are respected. However, this would be a hard commitment without clear guidance. Qualified intermediaries—such as EFSA—could provide guidelines on the implementation of these principles and, in turn, adopt them for their own activities.

Code of conducts represent the most viable way for the implementation. To date, machine learning and data-driven risk assessment do not represent the state of the art in food safety but rather an alternative and promising approach. This entails that on the one hand that strict legal requirements might undermine the exploration of these solutions; on the other hand, compliance to voluntary rules for their deployment can foster a trustworthy and responsible innovation.

Speaking of trust, the Blueprint could serve as an additional evaluation guideline for EU initiative regarding AI Trustworthiness, i.e. the ALTAI (The Assessment List on Trustworthy Artificial Intelligence).[1] This tool, released by the HLEG in July 2020, is intended to provide a self-evaluation instrument for organisations willing of assessing the trustworthiness of their AI tools by verifying their adequacy to the seven key requirements—Human Agency and Oversight; Technical Robustness and Safety; Privacy and Data Governance; Transparency; Diversity, Non-discrimination and Fairness; Societal and Environmental Well-being; Accountability—identified by the EU Commission and discussed above. It has been noted that some of these requirements are particularly significant in the domain at stake. Therefore, the Blueprint could help organisations in identifying, integrating or aligning their practices to the general requirements identified by the Commission. Vice versa, the Blueprint could be used by the Commission to highlight areas of particular risks from domain at stake (e.g. the conclusions and the replicability of studies relying on machine-learning) and act accordingly. Furthermore, the newly introduced CAPAI (Conducting Conformity Assessment of AI Systems) model[2] might constitute a valid point of integration for the Blueprint.

6.3 The Next Steps in Food Data Governance

Finally, let us observe some possible outcomes in food data governance and how research could contribute to understand them.

The technical premises of this book are based on the assumption—yet, supported by a concrete interest shown by EFSA, independent studies and commercial

[1] Available at https://futurium.ec.europa.eu/en/european-ai-alliance/pages/altai-assessment-list-trustworthy-artificial-intelligence, accessed on 27/04/2022

[2] Floridi et al. (2022)

entities—that an ongoing datafication trend can be observed at every level of food safety risk assessment, from information gathering to data analysis and knowledge extraction. The level of innovation in this field is evolving constantly, with occasional sudden peaks such as the automation of literature reviews. Therefore, research in data ownership and governance has to keep pace with cutting edge technologies deployed in this sector.

With the Transparency Regulation just entered into force, it will be necessary to revise the commentary provided in Chap. 2 in the light of EFSA's implementations. Undoubtedly, contributions from scholars active in food and environmental law will provide further insights on the new Regulation and its effectiveness. Still on the legal side, the relationship of our findings and Corporate Social Responsibility is undoubtedly an area of valuable investigation purposively neglected by this dissertation, as well as the topic 'Big Data and competition' which requires more investigation. Furthermore, with the food market becoming increasingly global and the progressive extension of the food chain, legal comparison is a research methodology that can become useful in evaluating how different jurisdictions deal with the risks posed by the ongoing datafication trend. Additional values and principles might also derived from other domains, including environment protection, animal rights and welfare, and sustainable development. Their integration can constitute both a validation and an enhancement of the Blueprint.

The proposed Data Governance Act and the AI Act represent a significant forthcoming challenge. The potential implementations of our Blueprint, as well as other principles emerging from AI ethics discussion, are numerous under these pieces of legalisation. However, their integration with other legal frameworks governing data must be established. For instance, how the Data Governance Act will interact in practice with the GDPR, the Regulation on Non-personal data, and the Copyright Directive (especially as regards exceptions for text and data mining) is an open question. The legal scenario would be further complicated by its interplay with the legislative framework applicable to food safety: the Transparency Regulation (and the GFLR), the Aarhus Regulation and Regulation 1049/2001. This interaction will reveal the degree of compatibility as regards access and distribution rules. To put it differently, the Data Governance Act might provide valuable opportunities to increase data sharing, but only the practice will reveal the extent to which they are viable within and compatible to the legal framework in place. Similarly, the final version of the AI Act will provide the legal basis for forms of self-assessed compliance for the use of machine learning systems. Other forms of implementations—yet, unlikely—would first require the framing of AI systems used in our domain within the legislative coverage of the Act, e.g., high or medium risks.

When discussing possible implementations, some proposals have been made, yet in a purely hypothetical form. More work is then needed to assess how this introductory framework could be translated into practice. Such translation can occur both at legal and technical level, hence the need of further research in both fields of study, as well as in their interplay following the *by design* approach. The cross-fertilisation of heterogeneous research areas seems unavoidable for any future work.

Reference

Floridi L, Holweg M, Taddeo M, Amaya Silva J, Mökander J, Wen Y (2022) capAI-A procedure for conducting conformity assessment of AI systems in line with the EU artificial intelligence act. Available at SSRN 4064091

The manufacturer's authorised representative in the EU is Springer Nature Customer Service Centre GmbH, Europaplatz 3, 69115 Heidelberg, Germany. If you have any concerns regarding our products, please contact ProductSafety@springernature.com

Printed and bound by CPI Group (UK) Ltd, Croydon, CR0 4YY
28/09/2026
02224071-0003